'This is the mos... ...kable
sea change that is taking place in our ... ...e nature of
human language.' – *Michael Corballis, author of* The Recursive Mind:
The Origins of Human Language, Thought, and Civilization *and
Professor Emeritus of Psychology at the University of Auckland*

'This is exciting work. I learned a tremendous amount from it, as will
anyone who is concerned with the nature of language and of mind.'
– *Robert Brandom, University of Pittsburgh Distinguished Professor of
Philosophy and Member of the American Academy of Arts and Sciences*

'Daniel Everett's book is a must-read for anyone with an interest in
knowing what makes us human … Everett resets the research agenda
for linguistics, psychology, and neuroscience.' – *Philip Lieberman, Fred
M. Seed Professor of Cognitive, Linguistic and Psychological Sciences
and Professor of Anthropology at Brown University*

'An impassioned argument that language has adaptively emerged
as our species' "tool" for achieving social collectivity via discourse.'
– *Michael Silverstein, Charles F. Grey Distinguished Service Professor of
Anthropology, of Linguistics and of Psychology, University of Chicago*

'*Language: The Cultural Tool* represents a radical reassessment of the
origin and evolution of language … The book eloquently reminds us
that the incredible diversity of languages on this planet reflect different
ways of thinking and being in the world – a phenomenon that might
sadly be on the verge of extinction.' – *Robert Greene, author of* The 48
Laws of Power, Art of Seduction, 33 Strategies of War, The 50th Law
*and* The Descent of Power

'Everett offers a provocative perspective on the questions of the nature
of language, of how language evolved, and of how languages are
structured. It is a significant contribution to the contemporary debate
surrounding these issues.' – *Robert Van Valin, Professor of Linguistics
at The University of Buffalo and Heinrich Heine University, Dusseldorf*

Daniel Everett was born in California. He lived for many years in the Amazon jungle and conducted research on over a dozen indigenous languages of Brazil. He has published on sound structure, grammar, meaning, culture and language. He is currently Dean of Arts and Sciences at Bentley University in Massachusetts. Previously, he was Chair of the department of languages, literatures and cultures at Illinois State University. He is the author of *Don't Sleep, There Are Snakes*.

# LANGUAGE:
## The Cultural Tool

ALSO BY DANIEL EVERETT

*Don't Sleep There Are Snakes*

# LANGUAGE:
## The Cultural Tool

Daniel Everett

PROFILE BOOKS

This paperback edition published in 2013

First published in Great Britain in 2012 by
PROFILE BOOKS LTD
3A Exmouth House
Pine Street
London EC1R 0JH
*www.profilebooks.com*

3 5 7 9 10 8 6 4 2

Typeset in Minion by MacGuru Ltd
*info@macguru.org.uk*

Printed and bound in Great Britain by
CPI Group (UK) Ltd., Croydon CR0 4YY

A CIP catalogue record for this book is available from the British Library.

ISBN 978 1 84668 268 1
eISBN 978 1 84765 415 1

To my mentors and former officemates,
Marcelo Dascal and John Searle,
rare teachers both

# Contents

'Instead of notions borrowed from books and immediately changed into philosophical concepts, I was confronted with the lived experience of native societies, by the commitment of the observer. My mind escaped from the claustrophobic steam bath to which it had been confined by the practice of philosophical reflection. Led to the open air, it felt refreshed by the new breath. Like a city-dweller released in the mountains, I became intoxicated, while my dazzled eyes examined the richness and variety of the scene.'

Claude Levi-Strauss, *Tristes Tropiques* (1955)

'If a man is offered a fact which goes against his instincts, he will scrutinize it closely, and unless the evidence is overwhelming, he will refuse to believe it. If, on the other hand, he is offered something which affords a reason for acting in accordance to his instincts, he will accept it even on the slightest evidence. The origin of myths is explained in this way.'

Bertrand Russell, *Proposed Roads to Freedom* (1918)

## Preface

People of the twenty-first century are developing new technologies that have already altered the foundations of learning, teaching, art, science, politics, government, business, music, and literature. The most interesting aspect of these exciting innovations is that they are all made possible by a single tool, human language, *instrumentum linguae*.

The idea that language is a tool has been around for a long while. Lev Vygotsky, the great Soviet psychologist, was one of the first to make this claim explicitly in modern times, though Aristotle framed language in these terms more than 2,300 years ago.

But no one has quite gotten around to weaving together the findings of modern linguistics, psychology, and anthropology to flesh out the meaning of the hypothesis that language is an artifact, a *cultural tool*. An instrument created by hominids to satisfy their social need for meaning and community. This is our ambitious project.

Some experts say language is an instinct, rather than the invention of a community of minds. Most, however, believe that instincts are simple, unlearned reflexes. An instinct is a baby's desire to suckle. But language is learned and complex, a work of function and form developed and honed by human cultures since the dawn of our species.

# THE GIFT OF PROMETHEUS

> *'Then Prometheus, in his perplexity as to what preservation he could devise for man, stole from Hephaestus and Athena wisdom in the arts together with fire – since by no means without fire could it be acquired or helpfully used by any – and he handed it there and then as a gift to man.'*
>
> Plato's Protagoras

The Greeks told a myth about one of mankind's greatest tools, fire. The story's hero was Prometheus, whose name means foreseer. Prometheus grew fond of the creatures that Zeus had asked him to help create, man and woman. He watched them with pity as they huddled cold and fearful of the dark, stumbling blindly after every setting of the sun. He knew the solution to their problem – fire. But Zeus did not want humans to have fire. Fire would give humans more power than Zeus intended. They might even rival the gods themselves. So Zeus forbad it.

Prometheus knew the risks of disobeying the king of the gods. Yet for pity and for love he smuggled a charcoal lit by Apollo's fiery chariot out of Olympus in a fennel stalk. No matter how pure his motives, Prometheus paid a horrible price for his charity. Zeus condemned him to an eternity of pain chained to a rock in the Caucasus, where each day his liver was consumed by a large vulture, regenerating every night in order to fuel his pain on the morrow. Only when the mighty Hercules slew the vulture and broke the chains was Prometheus freed.

The myth of Prometheus, like all good myths, encapsulates cultural values and offers answers to keep a group of *Homo curious* satisfied until a better answer comes along. In this myth we can take away the belief that fire originated once in the human story. We are given a glimpse of the problems that fire was meant to solve. And we are taught that the coming of fire was a momentous event in human history.

The Hebrews' myths also include a narrative about their gods coming to fear the growth of human power. But the Hebrew story differs dramatically from the Greeks'. The Hebrews' scriptures recognize that the power of language is greater than that of fire. The Hebrew god is not threatened by humans' control of fire, but rather by their ability to talk to one another. From this appreciation for the power of language emerges the Hebrew myth of the Tower of Babel – the tower that was raised to threaten the gates (Bab) of god (El). In this myth God is not worried about the physical technology of his creation, whether picks, axes, fire, or the like. He is instead infuriated by humans' ability to work together. This threatens his power. And their cooperation rests upon on their communication. So God scatters his people across the face of the earth. Or as the Bible puts it:

> And the LORD said, 'Behold, they are one people, and they all have the same language. And this is what they began to do, and now nothing which they purpose to do will be impossible for them.' 'Come, let Us go down and there confuse their language, that they may not understand one another's speech.' So the LORD scattered them abroad from there over the face of the whole earth; and they stopped building the city. Therefore its name was called Babel, because there the LORD confused the language of the whole earth; and from there the LORD scattered them abroad over the face of the whole earth.

Genesis 11: 6–9 New American Standard Bible

Ironically, the Hebrew god was not a linguist. He did not seem to realize that diversity strengthens *Homo sapiens*, and diversity in language and culture strengthens us the most. According to the Bible, God created one man, Adam, and gave him the charge of learning about and naming the flora and fauna of creation. By spreading Adam's descendants around the globe God in effect created a thousand Adams, learning about and naming not just the Garden of Eden, but the entire world

– wherever the children of Prometheus have gone, they have taken fire and language to master and learn about their world. This means that no one of us speaks the 'right' language. We all speak the language(s) that helps us and these languages are formed to meet the needs of our culture and social situation.

The Hebrews were right about one thing, though. The uttering of the first noun or verb, as non-momentous as that sounds, was arguably of greater importance than the stealing of fire from the gods of Olympus. Nouns and verbs are the basis of human civilization. Without these and other words, we could not utter history and life-changing phrases like 'I now pronounce you man and wife', 'This must be the place', or 'I name this ship the *Titanic*'. If it were not for words, Founding Father Patrick Henry could never have uttered his famous sequence of two nouns, one pronoun, one disjunctive particle, and one verb, 'Give me Liberty, or give me Death!' With nouns and verbs society was founded. With nouns and verbs the growth of human knowledge began.

Naturally, therefore, a research question that captivates many modern thinkers is precisely the origin of nouns, verbs, sentences, stories, and other elements of human language. Did language and its parts come about suddenly or did they emerge gradually as cultural adaptations?

This book is about the development of this great linguistic tool of our brains and communities, the cognitive fire that illuminates the lonely space between us far more brightly than the light of flames ever could. Here we look at the story of mankind's greatest tool, its purposes, and how it might have come to be.

Unlike physical fire, the cognitive fire of language did not exist before humans called it into being. And every individual and culture in the history of our race places its own mark upon this tool. It is an invention that envelops all humans. It unites. It divides. It warms our hearts. It chills our souls. It invigorates our bodies and steels young men for battle. It gives us the greatest pleasure of all – focused and ordered thoughts. We have become *Homo loquax*, as author Tom Wolfe calls us, or 'speaking man'. We are the masters of this raging cognitive fire.

Language's contribution to our mastery of the world is one way in which it serves as a tool. It is our greatest display of cognitive technology. It is the basis for an arsenal that includes mathematics, science, philosophy, art, and music. Language enables our brains to do things they could not do without it, like solving arithmetical problems, following recipes, and thinking about where our children are going after school.

No linguist, psychologist, anthropologist, or philosopher would disagree that language is useful. But there is enormous disagreement about where this tool came from. Some say that language was discovered by chance, like fire. Others believe that one brilliant *Homo sapiens* might have invented it 75,000 years or so ago, as the Cherokee chief Sequoya invented writing for his people. Still others claim that language is genetically encoded in the human mind, the fortuitous by-product of packing our skulls full of an unprecedented number of neurons.

Easily the most famous answer to this question, though, is that language is part of our genetic endowment and that, because of this, all human languages share an almost identical grammar –which includes sound systems and meanings. Under this view, the only significant differences between languages are their vocabularies. But this is not the only available explanation for the growth and presence of language in all humans. As I have said, I do not even think it is the best answer.

This is not a book about why one view of language is wrong and why another view is correct – although it does not shy away from stating its conclusions. Rather, this is a story about the joy of language, a joy that has filled my soul during more than thirty years of field research among indigenous societies of the Americas and life among my fellow *Homo loquaces*. From each of the nearly two dozen languages I have studied in the Amazon, Mexico, and the United States over the past decades, I have learned things about the nature of our species and our ability to communicate that I never would have learned by living a different life. I have learned about humans' relation to nature and about perspectives on living and speaking in a world delineated by the ancient cultures of the jungle. I have learned how words reach into my heart and change my life, from the poetry of e.e. cummings and the prose of William James to the fireside stories of the human family. Language gives humans their humanity.

But how did this marvelous artifact originate? How is it that all humans possess it? Why are there so many similarities between languages if each one is a tool for a specific culture? And what does it mean, finally, to say that language is a tool? Is this just a way of speaking?

The last question answers them all.

Most humans are fascinated by language, by our species' ability to talk, to inform, to persuade, to insult, to lie, and to praise – that is, to express the range of our thoughts and feelings through symbols in the form of sounds, gestures, marks on paper, drum beats, and the myriad of other ways we have found to use our senses and brains for communication. It is only natural that we should be so fascinated by this communicative technology, for nothing has more to tell us about what it means to be human than the forms, sources, and uses to which we put language. It is the foundation of every human advance, from Cro-Magnon cave paintings to Thomas Edison's light bulb to Mark Zuckerberg's Facebook. Upon it rests the 'information age.'

All human abilities, including language, derive from two sources – genes and environment. The idea that language is exclusively a product of our culture or social environment is as simplistic, unhelpful, and wrong as the opposite idea that language grows like hair, shaped by our genome with no significant learning involved. Language, like human reasoning itself, emerges at the nexus of our biological endowment and our environmental existence. The relative proportions of nature and nurture necessary for the creation of the language device are agonizingly difficult to determine.

Yet it is the recipe for the alloy of language that has become the eye of an intellectual storm surrounding theories of human communication. Linguists, psychologists, anthropologists, biologists, and philosophers tend to divide into those who believe that human biology is endowed with a language-dedicated genetic program and those who believe instead that human biology and the nature of the world provide general mechanisms that allow us the flexibility to acquire a large array of general skills and abilities, of which language is but one.

The former often refer to a 'language instinct' or a 'universal grammar' shared by all humans. The latter talk about learning languages as we learn many other skills, such as cooking, chess, and carpentry.

The argument that we possess a language instinct usually relies on two observations: the commonality of language – all humans speak languages; and the conformity of languages – all languages share highly specific features. For the alternative proposal, that language is invented and transmitted culturally, empirical support comes from the general wattage of human neurology, our capacity for learning, and from the knowledge that many properties of language are forged by forces outside the brain. This latter proposal takes seriously the idea that the function of language shapes its form. It recognizes the linguistic importance of the utilitarian forces radiating from the human necessity to communicate in order to survive.

The purpose of what follows is both to inform and to debate. Many of the matters I discuss are controversial. But this does not make them less worthy of consideration. One of my goals here is to give alternatives a fair hearing, even if I ultimately reject them to return to the safe harbor of majority opinion. However, I will need to make a case against one view in order to make a better case for another. This is standard scientific discourse. I believe that, so long as we are civil and our evidence is worthy, it is healthy to speak negatively of one matter and positively of others.

*Language: The Cultural Tool* explores one simple idea: that all human languages are tools. Tools to solve the twin problems of communication and social cohesion. Tools shaped by the distinctive pressures of their cultural niches – pressures that include cultural values and history and which in many cases account in many cases for the similarities and differences between languages.

First, though, we need to have a definition of both language and culture, in order to comprehend how they work together. Language is how we talk. Culture is how we live. Language includes grammar, stories, sounds, meaning, and signs. Culture is the set of values shared by a group and the relationship between these values, along with all the knowledge shared by a community of people, transmitted according to their traditions.

I have written this for the reader interested in the nature of human language, the system that binds all of us together, the tool that allows us to communicate outputs of our minds to others. My discussion is directed at thinking about the broadest and most important issues. But there are times when it is necessary to confront the details. We need both a wide-angle lens and a microscope to view the composition and the context of language.

There are things that we need to know to come to an informed opinion about the nature of language – its use, its functions, and its forms. One such is its origin. So the discussion that follows begins on the African savannah. That can help us understand how language solves problems – and how it kept our early ancestors alive and thriving. These problems must be confronted. We need to know how language is used by different societies. We have to understand what human bodies and brains must be like to produce language. We must examine the relationship between language and culture in numerous communities. And we need to see where the evidence lies for the conflicting ideas that language the tool is either innate or an invention.

As we get further into the issues, it will become clear that there is no unambiguous evidence for language being innate and that the very concept of 'innate' is too ambiguous to aid us in our quest to understand language.

But if language is a tool that is invented rather than an attribute of our genome, then the intriguing issue of similarity among the world's languages arises. Why do languages have so many features in common if these features are not part of the genome? And if we can answer this, an equally difficult question arises. How dissimilar can one language be from another?

The latter question can stir people's emotions. Even among many scientists it is supposed that all languages are 'equal' in some way. Most linguists, for example, will say that human languages are equally complex or that one language is as versatile as the next. Really? Are languages homogeneous then? But these stock answers confuse the equality of languages with the equality of human biology – human bodies and brains. I believe in the latter, but not in the former. The idea that 'all languages are created equal' seems grounded less in research than in a sense of political correctness. At times one gets the impression that if all languages do not share the same level of complexity or versatility throughout, then some languages must be inferior to others. People seem to worry that if we say a given language lacks grammatical devices that are found in other languages, then that this is tantamount to claiming that the speakers of one language are somehow inferior to the speakers of the other. But nothing could be further from the truth. I am not inferior to my son because he plays golf and therefore uses golf clubs, while I do not. I just don't need golf clubs. In much the same way, languages are tools that fit their cultural niche. From an evolutionary perspective, creatures use what they need and not what they don't. There is nothing politically incorrect in that.

At the same time, much of the emotion in the debate about relative language complexity is beside the point for the simple reason that we lack descriptions of too many of the world's languages to know what the 'standard' level of complexity is for human languages. In fact, no one has given or even proposed a coherent definition of 'linguistic complexity' that is accepted by all scientists, though there is research being conducted on this topic now in several parts of the world.

But for some, the boundaries of language are not to be discovered by empirical research but rather by mathematical analysis. It is not necessary to conduct field research to discover the axioms of set theory. If language were solely mathematically based, we would not need to conduct field work to discover its properties, rather we would need do nothing beyond the deep contemplation of our own native language. If one adheres to this omphaloskeptic conception and its implication that the properties of languages can be deduced from a general theory without the need for spadework, then the idea that language is a cultural artifact will make little sense. But the evidence points away from the idea that languages are derived from axioms. Growing evidence leads to the conclusion that the differences and similarities of the world's languages are outgrowths of neither mathematics nor genetics, though they have properties that are constrained by both.

This book is divided into four parts. In the first part, covering the material in chapters one through four, I examine the problems of communication and survival that confronted our evolutionary ancestors. I talk about the absence of specialization of human biology for language and show how our skills at general reasoning and the formation of human communities and social interaction might have shaped the emergence of human language.

In the second part of the book, from chapters five through eight, attention is focused on the solutions that nature and environment provided to the communication problem. I review the evidence that there are many solutions, not merely one, to the communication problem worldwide, considering the differences between languages of the world, as well as their similarities. I also look at the biological, mental, and socio-cultural platforms that are needed for humans to be capable of using, learning, and understanding languages. I delve deeply into the forms and functions of language, from sound structure to sentence structure and beyond. And I consider the single candidate for an instinct for language – Aristotle's proposed 'social instinct.'

Part three of the book – chapters nine and ten – looks in considerable detail at how the often invisible hand of culture shapes the forms and meanings of human grammars and languages.

In the fourth and final part of the book, my focus is on the diversity of solutions to the language problem, the importance of cultural and linguistic diversity to the survival of our species, and how human grammars can lead to happiness.

Many researchers are working today to understand how language and culture mold one another. The specific part of this question that I want to answer is how it is that the values that we hold as members of human societies shape the ways we communicate. Culture does not do it all, of course. But neither does the mind. Our cultures, linguistic forms, and minds evolve together from birth to death and even beyond the lifespan of any individual – each language is a history of the symbiosis of grammar, mind, and culture. This is why it threatens us all to see so many languages in the world threatened. The diversity of languages and their features set the perimeters of the human experience.

Language takes us through our human world. It is the theme of myths, philosophy, literature and science. In the vast majority of the world's literatures, both oral and written, humans have tried to explain the origin of their tools, abilities, and circumstances. In our early literary history, we used myths. Today we use science. Science is usually better than myths at explaining. But the myths arguably capture the grandeur of their subject better than science, because of the broad sweep of human emotions they portray, and the depth of their connection to the cultures from which they come.

# PART ONE:
Problems

## Chapter One
# LANGUAGE AS A SOCIAL TOOL

*'To say that truth is not out there is simply to say that where there
are no sentences there is no truth, that sentences are elements of
human languages, and that human languages are human creations.'*
Richard Rorty, *Contingency, Irony, and Solidarity* (1989)

Perspiring and exhausted, I stumbled through the Chiapan jungle looking for food. Ant bites had swollen into bulging, itching welts winding from my calves up to and very much including my crotch.

I was in this predicament in Mexico because I wanted to be a missionary in the Amazon. It was 1976 and I was training to be a member of Wycliffe Bible Translators. I wanted to translate the 'word of God', the Christian (Protestant) Bible, for a group of Amazonian Indians. I believed then that the Bible's words had the power to change lives for eternity. Part of the required training for this calling took my family and me to the jungles of Mexico, where the Mayans once ruled, for a four-month survival training course. There we learned to live and function in an extremely primitive and rustic environment. Most of the new knowledge and skills I was learning revolved around new tools – compass, knots, machete, mud stoves and other things I hadn't ever thought much about before.

My first task, after our initial few weeks of classes, was to build a house for my family in the jungle. I was prohibited from using nails, a hammer, or any other tools save my Swiss Army knife and a machete I purchased from the mission store. I constructed our 'home' from saplings and thatch from the jungle, tied together by jungle vines. I even made a bunk bed for my two daughters, Shannon (six years old) and Kristene (three years). I cut jungle wood to fashion a frame and supports for a mud stove. For the stove I used poles and vine to build a wooden box on posts I embedded into the ground. The contraption stood about three feet high. Then I filled it with wet clay and built the clay up into a chimney and a space in the base of the stove for firewood, covering it with a cast-iron cooking plate. Once I had my family set up and had purchased dry goods and a few live chickens from the mission to sustain them, I knew I would soon be called for the 'final exam' of all my jungle training – survival hike.

On survival hike, which all prospective members of the Wycliffe Bible Translators had to experience, male or female, you were called out by surprise, allowed to take only what you happened to have on you at the time (except food, which was confiscated), and left on your own for three to seven days in a part of the jungle you had never been to. You never knew how long you would be there. In anticipation of the hike, I kept a canteen full of boiled water, fifty feet of rope in a tight coil, a rain poncho, fish hooks and line, a pocket knife, my machete, and some matches on my person at all times.

On the day that I was called, I felt ready. Several of us stepped into a dugout canoe with an outboard motor on the back and sped off across Lake Ocotal. (Our location was about fifty miles east of the old Aztec settlement of Ocosingo, the nearest town. Ocosingo itself could only be reached by hiking out over the hills or flying above them from where we were.) On the opposite shore, we hiked several miles into the rain forest. From a certain point, the staff of the mission would look at one man and say, 'You stay here. We'll be back in a few days to get you.' My turn came and I was left as they walked on. I looked around me. Dense jungle. It was humid and very hot. I wanted to reconnoiter the area, find drinking water, and maybe gather some plants for dinner. But it was past 3 p.m. and I knew that night would fall suddenly around 6 p.m. I had to get to work. No time to worry about food. I would have to find water soon or dehydration would affect my ability to cope. So I began by finishing off the water in my canteen. There are many vines in the forest full of drinkable water. Next I successfully located the type of vine I had been trained to look for as it slinked down a tall tree in the jungle surrounding my campsite. I jumped as high as I could to cut it above my head, before cutting it again just above where it touched the ground. I knew that if you cut it next to the ground first, the vine would suck up all of its water higher into itself. I placed a finger over the bottom end of the vine and opened my canteen. I then removed my finger and quickly inserted the end of the vine into the top. One good cut like the one I had just made from the right vine can produce nearly a quart of water. I had enough to drink for a few hours. And the water was pure and fresh. No need to boil it or treat it. Just drink it cool from the jungle.

Having secured my water, I set about building a bed and a camp-fire. First I cut four sturdy poles which I stuck into the ground, at the corners of an imaginary rectangle that was to become my bed. I bent the poles into asymmetrical X-shapes, with the longest end downwards. Then I tied each pair of poles where they crossed with jungle vine. Next I notched each pole about an inch deep, two feet off the ground. Then, again with jungle vine, I tied cross poles at the level of the notches I had cut. Following this, I fitted longer, sturdier poles into the notches and secured them with vine. The frame of my bed was done.

Now I needed a mattress. So I cut several saplings, six to seven feet long. These were much greener and thinner than the poles I had used for the support posts. Laying down the green shafts, I took my rope and tied it to one of the long poles in the notches. I then looped it from there to the pole parallel to it, the loops all about three inches apart, until I got to the other end of the bed. Next, as though I were weaving, I worked the greener, more flexible poles up and over the rope, in and out of the loops I had made. When I finished, I had constructed, just as I had been taught by my survival teachers in the mission, a bed with a springy mattress. On one side of this bed I attached, again with vines, my rain poncho as a shelter. Then I went in search of firewood.

Gathering firewood is the most important task of the day, aside from finding water. The smoke from a fire, if the fire is built close enough to your sleeping platform, can keep mosquitoes from eating you alive while you try to sleep. The flames of the fire frighten away – in principle – most wild animals. You also need fire to cook your food (if you are fortunate enough to catch or find any). The fire gives you light at night. I spent hours cutting firewood. A base of dry wood and kindling is necessary to get the fire going. When it is hot enough, wet, green wood can be added to keep the fire going at a moderate pace, so that it burns neither too fast nor too slow. But while I was gathering firewood, I was unable to hunt, fish, or gather edible plants (my favorite and main food during this survival training was the fiddle-head fern, which is rich in nutrients). Teamwork, I realized, would have been crucial for my long-term survival. I could get by OK for a few days, but for a longer period I would not likely have made it without someone else there to help find food and, perhaps more importantly, talk to.

During that first night, I awoke to the sound of a large creature of some sort stepping heavily near my bed. I immediately looked at my fire and saw that I had let it burn down to embers during the night. I hopped up quickly and placed more wood on the still live coals, blowing fiercely through a little rubber tube with a copper ending, which I could place directly under the coals, that another missionary candidate had loaned me just prior to my departure. With this fire-oxygenating tool, the dry wood I had placed on the coals quickly burst into flames. I clutched my machete and looked around me. Hearing nothing else, I went back to sleep. The next morning, after sleeping moderately well under the circumstances, I awoke to a rumbling stomach and with a determination to find food, be it vegetable or animal. I knew I lacked the talent and experience ever to become a serious hunter or fisher, but surely I could kill enough to stay alive if I kept at it all day?

I walked away from my campsite deeper into the jungle, careful to mark a path with my machete as I went, one small strip of bark removed on the side away from camp and two small strips on the side of the tree closest to camp (the problem, of course, being that if I were to accidentally turn perpendicular to my path, I would not see the markings I had made on the trees). Hunger cleared my senses and after a couple of hours I saw a small brown blur run towards a pile of dead leaves. It stopped – furry and big-eared. I recognized it as an agouti.

Ambitiously, I raised my machete and, in an attempt at stealth, approached the rodent. It bolted just a few feet ahead of me. I waited a couple of seconds, then moved towards it. Again it bolted. 'I am never going to get close enough to this damned thing to kill it,' I despaired. For a moment I wondered if I could hit it if I threw my machete at it.

But then I realized the obvious: I could not risk losing my machete, my most valuable tool. As I tried to come up with an alternative plan, the sudden sound of cracking sticks and rustling leaves just off to my right startled me and broke my concentration. Another camper, Lloyd, broke out of the brush on my right, only a few feet from the agouti, even closer than me, and launched an arrow from a small bow. I turned towards the direction of the arrow and saw the agouti writhing on the ground. My colleague jumped for joy, laughing and yelling. He grabbed the still-moving dinner-to-be by the hind legs and turned towards his own camp, never noticing me standing in the jungle brush only a few feet away.

I made my way back to my little space, crestfallen. Along the way I stopped to pluck and eat my first food in two days, a handful of fiddle-head ferns.

A bow and arrow! Shit! Why didn't I think of that. (I hadn't even brought thinner rope to make one with.) Lloyd had planned ahead – he had brought light nylon rope, permitted by our rules, on his belt, to use as a bowstring. My survival depended on my ability to make tools to help me cope with my environment. And I had failed.

My friend and I shared a common problem: killing protein that moved faster than we did. He solved it. I didn't. Long term, Lloyd would surely have out-survived me in the jungle, unless I became more crea-tive in my tool-making. My jungle-camp colleague was hardly a pioneer in this predicament, though. Throughout our emergence as a species, human societies have faced this exact same 'protein mobility' puzzle. And for many millennia, the bow and arrow has been a common solu-tion to it.

How did this wonderful tool come about? There are several pos-sibilities. One idea is 'monogenesis,' a single (mono) origin (genesis). The bow might have been invented once and then shared by all the small bands of early humans before they left Africa. Evidence does suggest that a dozen or so millennia ago in Africa our ancestors were using bows and arrows. So the bow was probably invented even earlier. Because bows and arrows are made from biodegradable material for the most part (even today, wooden arrow heads and shafts are more common among Amazonian tribes than bone, stone, or metal heads), evidence for them in the fossil record is sketchy at best. Therefore it is possible that this invention, so crucial to human survival, might have been invented only once in human history. Maybe. Or perhaps several creative individuals independently invented the bow in response to the universal protein problem in various parts of Africa and the rest of the world. These questions get at the essence of what we are after.

It is possible that the impetus and general principles of bow construction were somehow made available by evolution through a specific configuration in the human genome. This could be the case and yet still be relatively recent. Or the basic genotype could be old but have been triggered into action suddenly and recently. It may have been the case that one day the growing number of connections between neurons in the evolving human brain crossed a threshold such that tool-making was a by-product of the new physics of the brain – suddenly, a special 'tool' organ or area of the brain grew to include hardwired knowledge for building bows when triggered by environmental problems to do so.

Our resources and time as researchers are limited. So it is important that we use care in selecting the hypothesis to investigate first. I choose the hypothesis that requires no unseen forces, whether gods or as-yet-undiscovered genes. I make this choice partially because the bow and arrow seems to be an obvious and relatively easy solution to the universal protein mobility problem. And it is a good idea to check out the most obvious solution first, for any problem. In this case, that would be the tool hypothesis. Therefore, I assume that the tool that enables me to halt mobile protein is the result neither of genes nor gods, but of human inventiveness.

But even though this might be the best hypothesis to begin with, it is not necessarily the most interesting one or the correct one. Nothing would be more exciting than to discover evidence that a god or alien showed an early human how to make a bow and arrow. Entirely unexpected information often produces the greatest learning experience. Instead of building our knowledge base up homeopathically, a little bit at a time, we make a giant leap. But leaps are rarer than baby steps, both in evolution and in learning.

It would be nearly as exciting to discover that there *was* information in the human genome or otherwise to be found in the infant brain for making bows and arrows. Now I accept the existence of the human genome, but I do not believe that it encodes information on how to make specific tools. Maybe I am wrong. If it happened that there *were* evidence for a bow and arrow complex of genetic or cortical material, I would not only be astounded by the wrongness of my world view, I would be all the more excited and delighted by the lesson learned. To understand why I am skeptical about a genetic origin for the bow, consider this: humans are remarkable for their relatively small number of genes; humans have fewer genes than corn. It is not clear what significance to attach to this gene differential. But it is not the possession of many highly specific genes that makes us smarter than a corncob. Rather, it is the symphony of these genes working together that makes human babies brighter than lovely yellow corn on harvest morning. Part of our strength as a species is that humans are more flexible and more variable than many other species. Our genes' effects may be less linear – that is, the connection between our genotype (the genes that we contain) and our appearance and behavior (our phenotype – the outworking of the genotype as it interacts with the environment) is less predictable, more likely to be the result of several genes working together, genes linked less tightly by the syntax of the human genome than the lock-step genes of corn and simpler organisms.

The point is that there is no evidence for strong matching between individual genes and complex components of human nature. In fact there is evidence that culture can affect genes, thus enriching the process of natural selection. Anthropologists Peter Richerson and Robert Boyd have made the case that 'the process of cultural evolution has played an active, leading role in the evolution of genes.' If they are correct, then culture affects our biological evolution and our genes. And our genes are the alphabet by which their syntax writes the outline of our lives.

The discussion of bows and arrows gets us to the central hypothesis of this book: that language is a tool and, like the bow and arrow, it was invented. It might have been invented in the course of human history once or multiple times. But if I am right, language is in the first instance a tool for thinking and communicating and, though it is based in human psychology, it is crucially shaped from human cultures. It is a cultural tool as well as a cognitive tool. There are many such tools, including the concept of heroes, scientific theories, and the wheel. And yet it seems clear that language is arguably the most important of all the utensils of our brain. Like the bows and arrow, fire, and other tools, it is part discovery, part invention. Languages are the imperfect outputs of the thinking of bipedal primates, refined gradually by the tasks they perform.

Language is a cobbled-together set of answers to different facets of the problems of communication and cooperation among humans. It might not even be the best tool for communication that one can imagine.

Language is complicated. Perhaps it is the most complicated and astounding invention in the history of our species. It might help us, therefore, before we turn to tackle the hardest problem of all, to look first at some other tools of our species. The set of technologies that are either cognitive or cultural or biological is vast. To explore the range of tools, I would like to look at a series of items that many might not consider tools, from a culture of central Brazil: a game tool, a culinary tool, a set of song tools, and some decorative tools.

In 2004, I received research funding to lead a team to the Xingu (sheen-GOO) Park of Brazil for the first-ever documentation of the Suyá (soo-YA) language. I learned after my team's arrival that the Suyá preferred to be called the Kīsedje (keen-SED-gee), their autodenomination, since Suyá is a term used by people outside their community. The Kīsedje language is spoken by approximately 500 people in the Xingu reservation.

Kīsedje is a member of the Gê linguistic family, which also includes the Kayapó (kai-ya-PA) and the Xavante (sha-VAN-chee), among others. Gê peoples are intimidatingly strong and vigorous – known for their ability to chase down wild game on an open plain, catch it, club it to death with a long hardwood war club, and carry it back miles away. They are so tall relative to other Brazilian Indians that when a Pirahã friend of mine from more than 800 miles away in the Amazon rain forest first saw a Gê Indian, he turned to me and asked indignantly, 'Why are we Pirahãs so short?' Gê peoples have become better known to the general public in recent years because the Kayapó chief, Raoni, toured with the rock singer Sting speaking about environmental threats to the Amazon.

After traveling to the Kĩsedjes' land I had to negotiate with the chief of the Kĩsedjes, Kuiussi (cu-yu-SEE), and his tribal council for permission to enter the official reservation and conduct research. Entering a tribal community is an honor and a privilege, carrying with it considerable responsibility. And most minority communities, like the Kĩsedjes, the Pirahãs (pee-da-HAN), the Banawás (ba-na-WA), and others, have reason to mistrust outsiders, based on the history of violence, exploitation, and domination that characterizes western relations with indigenous Americans over the centuries. Negotiations are always delicate. Kuiussi agreed to meet with me to discuss my research project at a churrascaria – barbeque restaurant – in Canarana, the nearest Brazilian town to his village.

Kuiussi is famous in recent history for his long association with the legendary twentieth-century Brazilian explorers Orlando, Cláudio, and Leonardo Villas-Bôas, founders of the Xingu Park, adventurers extraordinaire, and tireless servants on behalf of the Indians of Brazil. I knew that he was in his late sixties, but as he entered the restaurant at noon the next day, I was struck by how much younger he looked. His long and thick, obsidian-black hair rested on broad, brown, heavily muscled shoulders. Like other Kĩsedje men, the definition of Kuiussi's biceps was accented by armbands of white cotton. Most days he wears nylon gym shorts and flip-flops, showing abundant evidence of sinew and muscle from neck to toe, but today he was wearing jeans, a polo short, and tennis shoes in honor of the occasion.

I was also surprised by his retinue – at least fifteen Kĩsedje men, women, and children came with him to enjoy the free lunch the gringo had promised to all (I didn't know I had). Kuiussi heaped his plate high with food, then directed his attention to me, because my research associates had told him that I, the single, nearly pigmentless American in the restaurant, was the director of the project. Kuiussi assured me that he was indeed going to allow us to work with his people and that he would facilitate this, but that first he should let us know what the restrictions were.

One of the first things that all of the Kĩsedje men were interested in was whether I was going to profit financially from this project. They had the idea that I could become famous and make considerable money publishing books and articles about their language. They made it clear that any money to be generated from their collaboration belonged to them. Fair enough, I said, but they should realize that the money was negligible, beyond academic promotions or raises that I might receive from my work. These things were important to me, I admitted, but they were not going to produce significant community wealth on their own.

Kuiussi seemed skeptical of that assessment, but he did not pursue it. He was more interested in discussing some of his principal concerns about our conduct in his village. First, no sex between us and his people. He did not want outsiders stirring up emotions and disturbing his people's social lives by sexual contact of any kind, from flirtation to intercourse. Second, no pictures without permission. Third, no naked foreigners in the village. Kuisussi explained that, although we might see Kĩsedjes *sans* clothes, that was part of their culture. On the other hand, he stressed, public nudity is not part of the *cultura dos brancos* or 'pale people'. When whites go naked in public, he explained, they mean something very different by it than do his people. Kuiussi's understanding not only of cultural differences but of the very conceptual bases of culture itself struck me as deep and insightful. Indeed, he was right on many levels.

After talking for perhaps ten minutes, Kuiussi introduced me to his son-in-law, Nhokomberi (nyo-kom-BEAR-ee). Nhokomberi, known as Nhoko, owned a home in Canarana because he was a full-time employee of the FUNAI, an acronym for the National Indian Foundation of Brazil, and so spent large periods of time outside the village. We chatted about families and such and then returned to our highly calorific and cholesterol-rich food. After lunch, we agreed to meet again in the evening at Nhoko's house to further discuss the goals of my linguistic project. Nhoko also promised to show me photos of the Kĩsedje village in which we would be working, some 200 miles away.

Upon my arrival at Nhoko's home later that afternoon, his lovely young wife came up to me, smiling sweetly the entire time, with a necklace in her hands that she wished to place around my neck. The necklace was of carved black seeds and had an unusual, shiny, and star-shaped white centerpiece. I later learned that this object came from within the brain cavity of a species of fish known to many tribes exactly for this decorative bit of calcified cranial matter.

I thanked Nhoko's wife (Kuiussi's daughter) and we all walked into the house. At Kuiussi's request, I had brought several liters of cold soft drinks for the people to sip as we talked business. After greetings all round, we sat in chairs and nine or ten men – warriors in their demeanor and expression – faced me in a semicircle, where I sat in the hot seat in the middle. I was undergoing a rigorous background check.

My inquisitors seemed satisfied after thirty minutes of questioning and everyone abruptly lightened up and the men began joking among themselves as they walked into Nhoko's backyard to get more soda. Nhoko himself retrieved some photos from his bedroom to show me. He told me that he had a house and a wife in the city and one more of each in the village, known in Kĩsedje as Ngorotire (en-go-do-TEE-re).

The pictures included images of houses, a river, Kĩsedje men and women, and different kinds of Kĩsedje food, as well as miscellaneous dogs and other animals. One of the pictures that caught my attention was of a flat disk of manioc (cassava root) bread lying on the coals of a cooking fire. The bread was layered with meat. I asked what this dish was called in Kĩsedje, thinking to begin research for a Kĩsedje dictionary. Nhoko looked intently at the picture, as several other Kĩsedje men also turned their attention to the photo I had inquired about. Then they grinned as Nhoko responded in perfect Portuguese, 'Pizza!' The Kĩsedje had adapted a native dish, **beiju** – manioc bread – to imitate a dish that has made its way around the world. Pizza makes an excellent tool for delivering food that is high in calories and taste. It is a cultural artifact, a culinary tool. This calorie delivery utensil is so effective that today many cultures worldwide have pizzas.

A Kĩsedje tool that is on its way out is the lower-lip plate. Gê men are known for lip and ear piercings. The Kayapó and the Kĩsedje used to wear plates in their lower lips. Older men still do. The Xavante men still pierce their earlobes with small wooden stakes that they rarely remove. Each of these forms of bodily alteration/decoration has its own set of meanings, as anthropologists have discovered. Lip plates, ear piercings, and so on are cultural tools for communicating about the user's status.

A still thriving cultural tool is a clan-based competition used to promote group cohesion and clan pride. This tool is the famous 'log race' found in many communities of the Gê linguistic family, including the Kĩsedje. This event includes separate competitions for women against women and men against men. The people use logs that weigh in the neighborhood of 80–100 pounds for women and 170–200 pounds for men. The best analysis of a Gê log race comes from the leading authority on the Xavantes, the late David Mayberry-Lewis, former curator of South American ethnology for Harvard University's Peabody Museum. The village engaged in the race divides participants into two teams. Each team cuts a length of palm tree from a large log. Then a member of the team pulls a log on to his or her shoulder. Next, these log-carriers stagger down the trail, their team members running beside them, through the bushes, whooping and encouraging the person carrying the log. When a carrier tires or the team thinks it is wise, another team member offers a shoulder. By this relaying the two logs are brought back to the village and thumped down in the middle of the village. They are subsequently transformed into tools for a different purpose: forming the furniture for the men's meetings in the evenings.

Valuing toughness, village cohesion, and communication in a single activity, the log race is a notable social invention and tool, composed itself of various smaller tools (the log, the path, the village clearing, the village, and so on).

There are tools we use for a variety of purposes. Among the Kĩsedje there is even a kit of verbal tools that can be used to introduce the basic point that language is a tool, shaped in its form and function by culture. The most striking verbal tool kit of the Kĩsedje are their songs. Anthony Seeger's 2004 book, *Why Suyá Sing: A Musical Anthropology of an Amazonian People*, brings the significance of music as a Kĩsedje cultural tool to life eloquently:

> Kĩsedje singing creates euphoria out of silence, a village community out of a collection of houses, a socialized adult out of a boy, and contributes to the formation of ideas about time, space, and social identity.

The contribution of this study of Kĩsedje music is its view that music is an important cultural-cognitive tool for Kĩsedje society, used to perform tasks in building group cohesion and meaning.

The Kĩsedje themselves have their own explanation for the functions of each verbal form in Kĩsedje culture. These are: **kapérne**, used to refer to 'normal' speech; **sarén**, the label for speech that instructs (lessons from parents to children, from the chief to the community, from the shaman to the sick); **ngére**, referring to songs in general; and **sangére**, used for special songs and speech invocations at specific ceremonies. Each of these genres of verbal tools in Kĩsedje has subgenres that the Kĩsedje recognize and label by descriptive phrases. For example, if you ask about 'normal' speech, the Kĩsedje will ask whether you are referring to 'plaza speech,' 'bad speech,' 'angry speech,' or some other type of speech. If someone is speaking publicly, in the open plaza of the village, their speech has both a different form and function from other types of speech.

Each of these Kĩsedje genres can take various forms. For example, **sarén** was once used by young men who were about to move from their parents' home to the men's house for an initiation rite. Just before the ceremony began, the young man would learn a short, formulaic speech to recite at the appropriate time. Then, as the ceremony was under way, the celebrant would be decorated with bird down and ornaments. He would then *sing* a farewell in his mother's house, departing from the house slowly and dramatically as he made his way to the men's house. There he recited the instructional piece he was given, exhorting the entire village to prepare for the final ceremony. This sounds pretty familiar. We have similar rites in weddings and graduation ceremonies in all western cultures.

Each type of song and speech in Kĩsedje is shaped by its function – the purpose it serves in their society, its place of performance, and its audience and performer. Kĩsedje songs not only exist to fulfill social functions, but their very form is constrained by the functions they fulfill, such as the brevity of the young man's speech before the puberty ceremony – he must not delay the ritual, which applies often to many men at once, by a prolonged speech or song.

Modern western music is likewise shaped at least in part by its function. Rock 'n' roll, blues, jazz, classical and all other genres are recognized by their forms and functions. 'What does rock music do for me?' you might ask. Well, think about it. Rock does many things but it doesn't make you relax. Mainly it expresses energy, rebellion, and excitement. If we ask what any one genre is for, we are led to ask the same of others. So, what do the blues do? Classical music? These might not have obvious answers, but they are important questions.

I can illustrate this point here with the form and function of the blues, one of America's genuine contributions to the world.

First, what is the *form* of the blues? I am talking here about real, Blind Willie Johnson-Huddie (Leadbelly) Ledbetter blues. Here is a verse of my own composition to illustrate:

*My heart is going down the drain.*
*My heart is going down the drain.*
*Somebody please relieve my pain.*

All true blues is like this. The first line is sung twice, followed by a final line that provides a reason for sorrow, a plea, or some such. The music is equally formulaic. The first line of a verse is accompanied by a major chord (the 'tonic' chord). This line takes four bars of four beats each. The repeated version of the first line is then accompanied by a chord based on the fourth note of the scale – a C chord if the first chord was G, for example (this second chord is called the 'subdominant chord'). The second line is sung to two bars of the 'subdominant,' and followed by two more bars of the tonic. Finally, the last line, the plea, begins with two bars of the 'dominant' chord, corresponding to the fifth note of the scale, followed by two bars of the subdominant, and then ending with two bars of the tonic. That makes twelve bars in all. Hence the phrase 'twelve-bar blues.' Simple. But don't mess with the formula. Try to change it and you run the risk of ruining the blues, man. Somehow those twelve bars reach into your soul and paradoxically, since the words are about struggle and suffering, they make you happy. Blues masters from Robert Johnson to Joe Bonamassa have explored this small but beautiful musical land of the blues.

The formulae for producing lyrics and music for the blues (and rock and country – since both of these genres simply borrow a chord pattern that originated with the blues) are cultural products. The repetitive pattern of the blues is successful because it fits with what the blues do. It is simple, and its invention has been ratified by cultural success. The blues emerged from the painful lives of disenfranchised African-American slaves and their descendants. They sang them to express hurt, anger, and need, though in the singing they answered many of their own pleas and brought 'euphoria' out of their feelings of pain. The blues fits its context. The old taverns and homes in which the blues originated were places where people expressed their hardships in their music, in short pieces that were themselves simultaneously their own 'punchlines.' They had to be short. The blues are aphorisms of pain. Their structure broadly fits their themes – simple messages of woe and struggles put to simple, predictable, repetitive chord structures. Everyone who plays a guitar sings the blues at one time or another. You play the blues any time you want, but especially when you're sad. And from the blues have come the musical movements of jazz and rock 'n' roll that have changed every land that listens to them. As a teenager from a Middle Eastern country said recently on a BBC newscast: 'I have more in common with people who listen to the music I listen to than people who share my nationality or my religion.'

Blues and its jazz, country and western, and rock 'n' roll offshoots have become more popular at the same time that interest in classical music has declined. While there are many reasons for the ascendance of one and decline of the other, these reasons are all cultural, as members of their containing cultures choose one form over the other. Standards of attractiveness change. Standards of performance change. But the medium of consumption and transmission of music in modern society surely plays a role in shaping the form of music. The rise of the radio and television forced music to fit into two to five minute segments that occur between ever more frequent commercials. You can get Cream's version of Robert Johnson's 'Crossroad Blues' into that length of time, barring all the cool jamming on the *Wheels of Fire* album, but you cannot get Mozart's *Requiem Mass in D Minor* there. So Mozart goes to non-commercial stations that specialize in classical music for a dwindling audience and Cream went to the commercial broadcasters as contemporary artists like Lady Gaga do today. Form is determined by function.

Thus over time any culture's music, whether Kĩsedje or Californian, evolves in parallel with the cultural constraints it has to fit. One form of music takes shape and another form passes into disuse, just as with clothing fashion, architecture, food, automobiles, and so on.

Music, like language, like bows and arrows, and like the pizza pie, is a tool, shaped by the culture in which it appears. The human body's weight gain or loss can be a tool (consider the brilliant use of weight by Robert de Niro in his 1980 movie *Raging Bull*). Gaining weight can be a tool to avoid the military, as it was among some men of draft age during the Vietnam War. There is no question that culture affects the form of the body. For example, among the Pirahã people there is a wide variety in physical appearance among males. There are some with kinky hair, some with straight hair, some with African features, others manifest more oriental features. But if I take Pirahã men out of the village to town for medical help or some other reason, as I have many times, they all weigh the same (132 pounds, give or take a pound or so), are the same height (about 5 feet 2 inches), and all wear the same pant, shoe and shirt sizes. This is culture shaping the Pirahã body form, just as pioneering anthropologist Franz Boas claimed about the plasticity of the human form one hundred years ago when he measured immigrants to the USA and their descendants. The Pirahãs have an open gene pool, deriving from children born from 'trysts' with river tradesmen. So genes do not explain their homogeneous body form. This is a product of their culture. In fact, when Pirahã men stay in the city more than a month or so, they gain large amounts of weight – as much as twenty-five pounds in my experience. But within just a couple of months of their return to the village, they are back in 'standard Pirahã form.'

You might say, 'Well, OK then, the form of music, the body, and so on can be shaped by our culture, but how could this possibly be true of the less obvious mechanisms of language, such as its grammar, that are crucial factors in determining the forms a language takes?' This is a good question. Before I answer it, though, I need to clear up an ambiguity inherent in the tool idea. The ambiguity is this: a tool can be either a 'biological tool' or a 'cultural tool.' Obviously, language is a tool. No one would deny that. But is it biological or cultural? Arms, feet, legs, brains, eyes, and so on are biological tools. Music, bows, numbers, pizzas, frying pans, and golf clubs are cultural tools. Where does language fit? The dividing line rests on whether a tool is learned or just grows. Arms are not learned, they just grow. So they are biological tools. Numbers do not just grow, they are learned. So they are cultural tools. We do not learn to hear, but we learn to make music. So music is a cultural tool, but hearing is a biological tool.

In fact it seems so obvious to many non-specialists that language is learned that the question arises as to why anyone would believe otherwise. Well, some professional psychologists and linguists think that there are actually some very good reasons to believe that language is a biological tool rather than a cultural tool. I was persuaded by these reasons for a long while and many people find them convincing still, so we need to consider them carefully. It turns out that there are not many such reasons, so we can summarize them easily. And none of them holds up under scrutiny, though it takes some careful discussion to show this.

Here are the main reasons that have led some scientists to believe that language is a biological entity rather than a cultural tool:

First, it is claimed that children know things about language that they could not possibly have learned. In other words, that the structures and principles of the grammars underlying the languages they speak are not found and thus not learnable from the data that they are exposed to while they are becoming fluent little speakers. This is often referred to as 'Plato's Problem' (a term introduced by linguist Noam Chomsky) or as the 'poverty of stimulus' argument.

Second, people seem to acquire their native language with ease and with the same level of ability regardless of their intelligence (within normal ranges). This is not true, however, for cultural tools generally, such as rituals, music, theater, and the like.

Third, languages around the world share fundamental similarities that, so the argument goes, must result from a common core all humans are born with.

Fourth, new languages such as so-called 'creoles' often appear rather suddenly. And yet even when they do, they seem to manifest all of the properties that are independently recognized as universal by linguists, even though the children who produce these new languages are the first generation ever to speak them.

Fifth, children around the world go through all the same stages in learning their languages, as though they were guided by some in-built bioprogram in their language development.

Sixth, almost no one ever learns languages after puberty as well as they learn languages before puberty, suggesting that language must be learned before a specific critical period wired into humans.

Finally, the brain has clearly delineated areas that are specialized for language which are the same for all people in all societies. Genes, not culture, must be responsible for this.

These are the basic arguments that support the idea that language is innate, or an instinct. There is no uncontroversial evidence for any of them, as we will see in the following chapters. In every case, there is an alternative explanation that is based on learning and general cognitive abilities. The simplest hypothesis is just that language is learned, as we learn to use our spoons to eat pabulum or our hands to play a guitar.

Discussing the alternatives is interesting and worthwhile , however, because it forces us to better understand the human condition, to see that John Donne's phrase 'No man is an island' is both a lovely combination of words and a profound scientific truth.

## Chapter Two

# FROM FIRE TO COMMUNICATION

*'The world of appearances is complicated, and language has only verbalized a minuscule part of its potential, indefatigable combinations. Why not create a word, only one, for the converging perception of the cowbells announcing day's end and the sunset in the distance?'*

Jorge Luis Borges, 'Verbiage for Verses' (1926)

anguage is ubiquitous because it is a tool. But this does not tell us what it is. In order to talk about the nature of language, we need to answer the obvious question, 'What *is* language?' There are four distinct answers to this question worth considering. From these definitions, it is possible to take stock of what it is that needs to be explained about language the tool.

One simple definition comes from Merriam-Webster's online dictionary, which defines language as 'a systematic means of communicating ideas or feelings by the use of conventionalized signs, sounds, gestures, or marks having understood meanings.'

A second definition was suggested by one of my favorite personages, Henry Sweet, the great British phonetician and linguistic pioneer of the late nineteenth century, who served as part of the inspiration for the character Henry Higgins in George Bernard Shaw's 1912 play *Pygmalion* and for the subsequent 1964 movie *My Fair Lady*. Sweet defined language as 'the expression of ideas by means of speech-sounds combined into words. Words are combined into sentences, this combination answering to that of ideas into thoughts.'

An alternative definition was proffered by Bernard Bloch and George L. Trager, two important American linguists of the first half of the twentieth century. They suggested the following: 'A language is a system of arbitrary vocal symbols by means of which a social group cooperates.'

And then finally, we should not neglect one of the most influential definitions of language, that of Professor Emeritus at the Department of Linguistics and Philosophy at the Massachusetts Institute of Technology (MIT), Noam Chomsky: 'A formal language is a (usually infinite) set of sequences of symbols (such sequences are "strings") constructed by applying production rules to another sequence of symbols which initially contains just the start symbol.' That goes over just about everyone's head the first time they hear it, including mine, so an example of what Chomsky means by a 'formal language' is probably necessary to understand it. And we also need to understand Chomsky's definition because no individual in history has had greater intellectual impact on the study of language than Chomsky.

To do so, we can invent a language of our own, beginning with a miniature dictionary. You cannot have a language without words, so any theory of grammar will have to include words. In our new language there are only going to be three words: 'John,' 'Bill,' and 'sees' (it is a very small language). We need to know how to put our words together, though, so there are also going to be four rules of grammar.

In formal grammars, sentences are the biggest units. So descending from sentences down to the words they are composed of we begin with the rule that 'S,' a sentence, is made up of a noun phrase ('NP') and a verb phrase ('VP'). In linguistic notation, this rule would look as follows: S → NP VP. Next we need to say how noun phrases and verb phrases are made. So a second rule could be one that tells us that there are only two possible noun phrases in this language: 'John' and 'Bill.' This rule, again in formal linguistic notation, would be: NP → 'John,' 'Bill.' Knowing how to make sentences and noun phrases is not enough, though. We also need to know how to incorporate verbs into our language. Verbs are found in their own phrase, the verb phrase. So the third rule in our little artificial grammar would look like this: VP → V NP. This rule tells us that a verb phrase contains a verb and a noun phrase (the object of the verb). All we need now is a rule that tells us what the verbs in the language are. In this impoverished toy language, there is just one, given by our fourth rule as: V → 'sees.'

This language's grammar is small, but it is nevertheless properly constructed. This type of formal toy grammar is commonly used by many linguists, computer scientists, philosophers, and psychologists to test ideas. It is said that the rules just given 'generate' (or describe) the sentences 'John sees Bill' and 'Bill sees John.' In a formal grammar (concerned nearly exclusively with form), we do not care what the sentences actually mean. According to Chomsky, meaning is secondary to grammar and all we need to understand of a formal grammar is that if we follow the rules and combine the symbols properly, then the sentences generated are grammatical.

Each of the definitions of language discussed so far is incomplete in some way. But this is only natural because language has myriad facets and no single definition could or should capture them all. Even so, there is one definition that seems to work better than the others – Merriam-Webster's. It is easily the most useful, complete, and accurate.

Henry Sweet's definition falls short because it would exclude sign languages as languages (it would also exclude computer languages). Languages are not just made up of sounds. I also think that his emphasis on words is too narrow. All words are 'signs' (a meaning paired with a form), but not all signs are words or sentences (stop signs, 'thumbs up,' and other gestures come to mind).

Bloch and Trager's definition is inadequate altogether if we want our definition to be only of human languages. It would mean that baboons have language, since a baboon troop's grunts would fall under such a definition.

Chomsky's formal definition is most popular among many of those who are interested in the theory of grammar. The question these theoreticians want to answer is: How much of language can we explain by mechanics alone, with no reference to meaning? And yet to define language this way is like trying to define music only in terms of notes and rules for combining them, with no concern for the aesthetics of music, how it emerges from different cultures and different times for different ends, and why we like some arrangements of notes but not others (for example, why do certain chord progressions create emotional reactions and stand out above others)? One website observes with regard to the Beatles that 'serious music critics dubbed them as "fresh and euphonious" for their ability to switch from C major to A flat major,' an assertion which fails utterly to express the originality and effect of this chord change. Clearly the question of what makes music 'beautiful' or 'meaningful' can get pretty touchy-feely, so I can understand why people are attracted to definitions of language and music that avoid the issues of meaning and beauty. And this formal definition of language is of considerable interest because we would learn a lot about the human mind if it turned out that language, or even a subcomponent of it – grammar – could be reduced to a list of principles governing the combinations of words into phrases and phrases into sentences. It is indeed the case that the constraints on humans' manipulation, concatenation (stringing things, like symbols, together), and other effects of rules for linguistic forms are crucial components of understanding language. After all, from some vantage points – like that of the field linguist working away in some distant place on a language few have studied – knowing about 'linguistic mechanics' is essential. Writing rules, figuring out verb phrases, noun phrases, and so on in a variety of languages has familiarized me with the kinds of structures one can expect to encounter in natural languages.

For example, if I find that adjectives follow nouns in some language I have never studied before, then I know that it is likely that the verb will precede the object. I love finding this kind of correlation precisely because it seems so weird! Why on earth should the placement of adjectives relative to nouns have anything to do with the positions of the verb and its direct object? This just seems so damned random. But scientists know that finding links between apparently disparate facts can indicate that there is a deeper principle in play. Seemingly unrelated facts like this make linguistics interesting because they remind us to ask why the world should be the way that it is. That is one of the fundamental questions of science. We will not answer the question of why language is the way it is by focusing on the form of language. Rather, our main insights into language will come from a better understanding of meaning, culture, and history and how these each in turn shape the form of language. Any answers we can find along the way will tell us useful things about how the human brain works and how it got to be the way that it is.

Ultimately, the definition of language you or I find most useful will depend on where our interests lie. This is true of most intellectual endeavors. There are some language researchers whose interest is in the formal, meaning-free properties of language. There is a lot to discover and there are many potential applications of that knowledge in computer science and elsewhere. But it is unlikely that this focus will get us to the underlying nature of human language.

If we want to understand language as the intersection of culture, cognition, and communication, then none of the other definitions I have discussed will do it justice. The formula that summarizes my own concept of language is: Cognition + Culture + Communication = Language. This means that each normal human being has a brain, belongs to a community with values, and needs to communicate, and the confluence of these states results in language. Communication sets up a problem to be solved. Our brains solve it. But let us not forget that this communication problem does not need to be solved *ex nihilo* by each individual. Part of the benefit of belonging to a community is having a language to learn, one that others have developed for us. Our brains are up to this learning task because, through thousands of years of human evolution and the history of language, language and humans have grown to fit one another. That is, every society's values have helped mold a language to its needs over hundreds or even thousands of years.

The crucial components that distinguish my adaptation of the Merriam-Webster definition from the others mentioned above are culture and communication. Although just about everything in this book is relevant to each of these definitions, the definition that I want to work with will always be one that includes culture, general human cognition, and the need to communicate.

Now I know that definitions can often be a matter of taste and that they rarely offer a resolution to the interminable discussions they provoke. If you prefer another definition, fine. But remember that our definitions specify our targets. If we are after different aspects of a broader phenomenon, our definitions will not be the same. The definitions I have suggested do not exhaust the possibilities. They simply help anchor the discussion.

Armed with a working definition of language, then, we can ask a very interesting question. How might language have arisen in our species in the first place? No one really knows the whole story here. But an imaginative trip to the distant past might be worthwhile in order to approximate an understanding of what language arose to be a tool for and, in that context, how it might have begun.

It would be very helpful to find direct fossil evidence for language evolution. Too bad that is impossible. We can find bits of indirect fossil evidence for the evolution of human speech, such as drawings on cave walls or prehistoric tool kits that indicate a level of thought requiring societal cooperation and culture. This likely indicates that ancient artists and tool-builders had language. We can also look for prehistoric evidence for language by comparing the relative development of the vocal apparatus in Neanderthal fossils and modern humans. In any case, it must be that language in some form or another existed prior to the evolution of speech in sounds and gestures. It could have had the form of whistling, humming, or tones used in conjunction with a vowel inventory much smaller than that now available to *Homo sapiens*. But sounds and gestures leave nothing at all in the fossil record.

What about genes? Can we look back in time and discover the origin of language by finding when a specific gene might have entered the hominid lines? No. Not only is this not possible now but, no matter what you hear in the press from time to time, a 'language gene' is never going to be discovered. Language derives from a complex set of emotions, behaviors, and thinking patterns. No one gene could underwrite all of this.

After all, if we were looking for a language gene, what would we expect it to do? Create grammar? But grammar is just one part of language, and beings with just a grammar would not have language. In fact, we know that meaning drives most, if not all, of grammar – we communicate to get a meaning from our heads to someone else's – and meaning would have to appear *at least* as early in the evolution of language as grammar. There are several recent studies that make this point exactly – *The Origins of Meaning* (2007) by James R. Hurford and *The Origin of Concepts* (2009) by Susan Carey are two of the most important. Other recent books take up other issues on the origin and purpose of language. Studies like *Why We Cooperate* (2009) by Michael Tomasello and *Mothers and Others: The Evolutionary Origins of Multiple Understanding* (2009) by Sarah Blaffer Hrdy make the case that what is really unique about human communication is that we want to talk to each other in the first place. The evidence for this innovative idea comes from comparing humans with other primates and animal species.

It might therefore make sense to search for a gene responsible for our desire to interact with others of our species. Now, I agree that interaction is the necessary first step in language, but interaction is not language. There seems to be something akin to an 'interactional instinct' in our species and this needs to be discussed. But the instinct to communicate, or interact, precedes language and grammar. It does not explain them.

But we could be looking too far ahead on the horizon. Maybe there are genes for parts of language and the sum total of these genes is what some call 'universal grammar.' A linguist might ask if there is a gene for pitch control, one that manages the frequency of the vibrations of the vocal cords in speech production. Say that such a gene was found. It would not tell us much because, although pitch is certainly a part of language, it is important to many other human behaviors as well, the most obvious being music. Frequency also helps identify emotion. Nearly every single component of language multitasks – it can be used for other activities besides language, such as biting (teeth), licking (tongue), ordering our activities (sequential motor skill portion of the brain, also responsible for syntax), and so on. We could summarize the relevance of genes to language by saying that, just as genes in humans are not linked in some rigid and predictable fashion to human behavior, neither is the output of all our communicative components predictable from the nature of the individual components of communication. Language is the complex interaction of culture, lots of different types of cognition, the need to communicate, the nature of communication, grammar, human sound production, and so on. It is our most complex behavior. Finding a language gene is much less likely than finding an architecture gene. Any genetic account of language is only going to get us at best halfway to an understanding. To borrow a phrase from Beatle George Harrison, language is 'within you and without you.'

With all of this in mind, let us try to imagine how language, prior to speech, might have arisen in our evolutionary ancestors. Here is one possible scenario, borrowing occasionally from Frances D. Burton's excellent 2009 book *Fire: The Spark That Ignited Human Evolution*.

On a cold dark night about two million years ago, an ambling ape approached a fire of brightly burning grass and small bushes started by a lightning storm that had just passed over its part of the African savannah, in the area we know today as Koobi Fora, Kenya. The small, four-foot-high, upright-walking creature, whose descendants were to become known as *Homo sapiens*, overcame its natural aversion to fire because it remembered that the edges of fires were a good source of food. There were dead insects and other animals in the blackened earth behind a moving brush fire or at the outer ring of a stationary fire. The insects there were tasty and filling, if it could find enough of them.

The ape benefited from the vitamin B12 and protein of the insects, which were easier to digest after being cooked by the fire. In addition, it consumed bits of charcoal, ash, and earth adhering to its insect repast. The charcoal and encrusted dirt helped the creature to digest and thus incorporate into its diet materials it would not otherwise have eaten, but which were sources of necessary minerals, proteins, and vitamins. It might even have enjoyed for its own sake the crunch and taste of the charcoal and earth, as much as some modern traditional communities like the !San, or bushmen, of the Kalahari seem to do, when they bake ostrich eggs by pouring them from their shells directly on to hot coals and ash.

As it contentedly gorged on insects near the fire, our ancestor might have heard the approach of a predator – say, a large leopard. Startled, it may well have recoiled in fear as the deadly cat emerged from the darkness of the night into the light of the fire, snarling and growling, circling slowly and deliberately as it prepared to pounce on the hapless, soft and crunchy biped. The terrified primate might have flailed about, vainly attempting to scare off this menace from the dark. It could have thrown earth and grass ineffectively at the leopard. But then, ever focused on the predator, perhaps it reached down and grabbed something hard next to its feet – a brand from the fire, say – and hurled that, too. The unwelcome guest would have likely jumped back in fear. If anything like this happened, the hominid would have had a 'Eureka!' moment.

The small ape might have then expressed its recognition of discovery by throwing another flaming stick at the leopard. Again the feline would have jumped back and snarled. If it did not attack, our ape would have repeated the successful throw, quickly. And another. And another – until the cat slinked from the circle of light backed into the darkness, abandoning the field in search of less threatening prey.

Although this is speculation, we know that *something* happened which led to our early ancestors' mastery of fire and thus to fire's subsequent effects on human development. Defense almost certainly was an early use to which fire was put.

Whatever enabled our evolutionary ancestors to control fire for the first time, they would have not forgotten its benefits. Fire brought light from darkness. It replaced cold with warmth. It made inedible food edible. And it was a weapon! With fire our primate ancestors were able to scare off many of their predators.

The primate who discovered fire, whether it did so before any other member of its species or merely independently of any other, would have no doubt reacted as some of us do with a new tool – it would have wanted to show it off. Maybe he or she ran over and threw firebrands at its fellow apes, enjoying, perhaps laughing at, their fear as they jumped to avoid the flaming wood.

Almost certainly, apes were learning independently and simultaneously about the benefits of fire through their own observations of its effects on danger, hunger, and cold. Fire became common, even necessary. Every one of these apes had to have it. Maybe some learned that they could start new fires if they carried glowing embers in a handful of dirt, or burning sticks to another stand of dry grass. The fire technology spread. People became addicted to it. Entertainment, hearth, and weapon all at the same time. It was nothing less than the greatest technological discovery in hominid history.

According to Burton, and to Richard Wrangham in his 2009 book *Catching Fire: How Cooking Made Us Human,* as these ancestor apes and their offspring learned more about fire and lived with it, they began to transmogrify – a process that would, after a few million years, produce modern humans, *Homo sapiens sapiens.* From those apes to now a new species emerged –culturally, physiologically, cognitively. Physiologically, their descendants would no longer need the heavy teeth and jaw structure to eat raw flesh and whole cellulose. Instead, they could enjoy cooked meat and plants, which were much easier both to chew and to digest. Their teeth and jaws grew smaller. Culturally, they began to develop a sense of community during nights around the fire with family and others they recognized. Cognitively, their interest in, need for, and ability to communicate accelerated as they sat together in the evenings, huddling against each other for warmth, and then perhaps sleeping in a jumble of bodies in the ashes as the fire died down, just as many people – like the Pirahãs of the Amazon – do still today. The hearth preceded the home in developing the sense of belonging and family.

This combination of a desire for fire and the need for group assistance in maintaining fire would have led to greater social cohesion among these pre-humans, a sense of 'hearth,' and a sharing of labor in using and maintaining a new technology, just as it still does for modern hunter-gatherer communities. The primates' fire-strengthened sense of community increased as they learned more applications of cooperation and togetherness, while their sense of belonging to that community also intensified and became more conscious.

But fire was a harbinger of even greater things. It seems likely that another, even more monumental discovery would have followed closely (in evolutionary terms) after the discovery of fire – the recognition of other minds, coming on the heels of greater time spent communally around the fire. As early hominids socialized, something about them – as opposed to all other apes – led them to the astonishing realization that their fellow apes had minds like their own. With that recognition, these apes – or their descendants – were on their way towards the belief that their fellow hominids had similar goals, pains, and needs. And with this belief, they began to attempt to communicate the contents of their minds. The belief that others think like us leads us to what researchers call a 'theory of mind.' This is a vital step towards language, because if others of my species have minds that work like mine, then it is much easier for me to predict what they are likely to do in a particular situation, based on my understanding of my own mind's activity. Some psychologists call the application of the theory of mind 'mind reading.'

Beliefs about other minds do not require linguistic thought, however. Animals also have beliefs, arguably. And animals seem to be able to predict the behavior of other animals, such as their owners, on occasion. It is not actually required that early hominids had thoughts like

'Gee, John is just like me' but they must have recognized that the featherless bipeds they interacted with were different from the other animals. Experimentation has shown that some non-human animals might have a theory of mind. What is required to have such a 'theory' is to treat other animals as though you can understand their behavior. I would not be surprised to discover that some dogs or primates were capable of this.

We all mind-read. We are not always accurate in our readings, but we are surprisingly good at this none the less. Mind-reading is an important step towards language. If other people have minds like mine, then they might think like I do. If they think like I do, then maybe I can express the contents of my mind to their minds. If you saw me looking at a new Steve Vai Ibanez guitar in the window of a music store, you wouldn't find it terribly difficult to read my mind. Or if you watched me running towards a bus stop. Or pouring cereal into my bowl at 6 a.m. on a Monday morning – you would know that I intended to eat it. Minds are often easy to read like this. Mind-reading, on the other hand, is not merely predicting another's behavior. Lions can often predict their prey's behavior. Dogs know when their master is getting ready to take them for a walk. Even fish in a bowl in some cases appear to know when they are going to be fed. But we use mind-reading not merely to predict behavior but to explain it and to reason with those whose minds we are reading. There is no uncontroversial evidence that any other species has discovered that individual members of that species have minds that are alike, except *Homo sapiens*. In retrospect it seems so simple – just realizing that my fellow creatures have minds like I do and they think like I do. But such a realization was evolutionarily transformative.

This theory of mind provided the foundation for the hominids' first joint ventures involving planning and cooperation. A first such endeavor might have concerned fire itself – learning more about it, keeping it going, transporting it. The ancestors of modern humans came to depend on fire too much to leave it to chance encounters with the vestiges of celestial electrical charges. They wanted to start their own fires and to take fire with them when they traveled to other campsites. To do this, though, they had to work together. Some would have transported the embers in their hands on a mound of dirt, perhaps, while others carried kindling. Eventually some among them learned to make fire from rubbing dry sticks together, while still others gathered wood to burn. Fire continues to be treated as a social resource in similar ways in many societies. Culture followed fire with the transformative power of shared values and knowledge. Could genetic evolution have followed?

Modern examples of culture changing genes are not hard to find. Consider lactose tolerance. All humans are born with the ability to digest lactose, an enzyme of mother's milk. But for most populations, the ability to tolerate/digest lactose 'turns off' after early childhood. However, in some societies that raise cattle as part of their culture and include milk in the diet of adults, the genetically regulated tolerance of lactose is extended from childhood throughout adulthood. Several researchers argue that this genetic innovation is both very recent (c. 8000 years) and driven by a culture–gene interaction. This has interesting implications for our consideration of language, one of which is that the absence of other culture–gene mutations affecting specific languages of the world presents a problem for theories that propose a language instinct or universal grammar wired into our genes. Why should we all have the same instinct, given how rapidly changes in our genes can be produced by changes in our culture? I will look into such questions later. But for now the point is that both genes and culture can operate independently or together and that these are fundamental components of the human experience, extending even to human biology, when that biology results from cultural innovations, or, to put it in Darwinian terms, culture creates new selective pressures on human evolution. Human evolution has not stopped! It is always on-going. In fact, with the constant coining of new words and subtle changes of grammar, language evolution continues today.

Given the possible scenario sketched above, it is possible to suggest an answer to the question: Why is fire universal? Fire is universal because all people need fire and because it is within humans' capacity to invent or discover and to control.

That same reasoning could also explain why language is universal. Language and communication are vital to the survival of our species and so are universal. Both result from group relations. The root of the word communication is commune, a community. The group logically precedes the interactions within a group. Nothing too hard to grasp in that. But this idea does not preclude the possibility that language is useful for more than communication. It is clearly vital to bring order to our thoughts. However, our thoughts rely on meaning and meaning in the human sense derives mainly from community and culture. So the conclusion seems inescapable that humans developed community prior to communication, as the little imaginary scene above depicts. This idea is supported by the social organization of other primates, many of which, such as baboons, chimpanzees, and gorillas, have social communities in the absence of any but the most rudimentary forms of communication. Community precedes language. We share such activities with all other apes. But with the discovery of the theory of mind by our species, in its need to interact and cooperate (unique human features discussed at length in much recent research by primatologists, psychologists, and others), an important threshold of communication was crossed.

Momentous first steps towards communication began with the new communal experiences around the fire. Communication might have begun with a few grunts and notes hummed back and forth to express approval or disapproval by one ape of another's conduct. Perhaps these were understood and enjoyed, leading eventually to humor – grunting, laughing, or humming when someone fell over, for example – and otherwise worked their way into the nightly routine, joining the gestures and facial expressions the group of primates, as primates in good standing, had used since long before. As these apes evolved, their sounds, embedded in ever more closely knit communities, became more frequent even as the apes' desire to communicate strengthened.

As their repertoire grew, perhaps to include whistling, yelling with different pitches (as in some modern-day tone languages of the Americas and elsewhere), they no doubt began to spread beyond the social context of fireside chats. But language would be less useful if it were only used around the fire. What would these creatures of the African savannah have on their minds to talk about most of the time? I do not think that the answer to that is too hard to figure out – food, in all its forms, live and running, dead and heavy, hot and juicy. The first topics of conversation were unlikely to have been the lovely colors of the forest, the way that one of the apes cut its hair, or the fear of god. This all came later. The first topic, the eternal topic, was and is food. Where to get it. How to kill it. How to carry it. How to prepare it. Some of the communication about food would have been spontaneous, such as taking advantage of the fortuitous discovery of a wild boar on the path. 'Let's get it!' This would have been an important meaning to convey. And it could have been transmitted initially by little more than a grunt accompanied by vigorous pointing.

What do you need to be able to talk about to plan a group hunting effort? Well, a couple of main ingredients – words for things, words for things that things do, and words for things that happen to things. That is, we all need to have words for objects and the events and states that objects can be involved in.

Think of food-getting again. You cannot say, 'Grab it!' until you know what an 'it' is and how 'it' differs from things like 'grab,' and until you have a way of communicating the differences between 'it' and 'grab' clearly to others in the community. These are not easy conceptual hurdles. And even after these first barriers have been cleared, a higher one looms – acquiring the ability to make sounds, gestures and shapes etc. that our fellow creatures and fire-mates can recognize as referring to 'grab' and 'it,' that is, inventing and sharing *signs*, perceivable and distinct units reliably associated with a unique meaning.

Planning group hunts was almost certainly fundamental to the development of human language. The creatures in which language began to develop might have preceded *Homo sapiens* on the evolutionary tree. But all creatures prior to *Homo sapiens* appear to have lacked the well-developed vocal apparatus that modern humans have, though there is some speculation that *Homo neanderthalensis* had speech in spite of its different vocal tract.

If other creatures had language, therefore, they lacked speech as we know it. They would have been forced to 'talk without speaking' (as singer Paul Simon excoriates in 'The Sounds of Silence'). This is a possible scenario. And it requires no speculation about the unobservable past. There are modern-day examples of *Homo sapiens* who communicate without consonants and vowels, illustrating clearly the possibility that language could have developed before speech. The many current populations that use whistle speech also illustrate the use of language without speech, as do all sign language communities. Whether or not language preceded speech, speech and language would have affected each other throughout human evolution. These examples show, however, that speech and language are logically distinct. Whatever type of language tool we use – whistle speech, consonants and vowels, sign language, and so on – the task is to communicate and the tools must fit the task.

The task of communication was a daunting one for evolution to solve. But clearly not impossible since we do indeed communicate. Like any complex job, however, communication can be better understood by breaking it down into its component parts and asking what would be needed to solve each sub-problem of the whole task. We need to think, in other words, about what it took for our entire species to evolve from grunts to gab, from roars to rap. What did it take for us to cross the communication threshold?

## Chapter Three

# CROSSING THE
# COMMUNICATION THRESHOLD

'[I]t seems to me appropriate to view the ... features of natural languages and their use as "cognitive technologies."'

Marcelo Dascal, 'Language as a Cognitive Technology' (2002)

One day during my field research in the Amazon, I asked some
Pirahã men if I could go hunting with them. They agreed to
take me. We left early the next morning, walking quickly,
paying little notice to the undergrowth around the village. After we had
gone perhaps a mile or two into the jungle, the undergrowth disap-
peared, the trees grew larger, the air cooler and the light much rarer
as we found ourselves under the canopy of the ancient Amazonian
climax forest. As they sensed game, the men stopped speaking – or so
I thought at first. They were whistling to one another. At the end of
the day, as we made our way back to the village with our spoils of a
couple of birds and monkeys, I asked the men about the whistling. 'It is
"high-speaking"', they said. '*Xapógáópisai.*' Later I recorded a couple of
men whistle-speaking and, with their help, translated what they were
saying. I discovered that in Pirahã, as in many other languages around
the world, it is possible to communicate about any topic in any context
without normal speech, merely by the use of patterns of pitch, loudness, and length. You can recount jokes or lie, talk about the hunt, ask
about the family or tell tall tales – all by whistling. In addition to whistle
speech, the Pirahãs have hum speech, another form of communication
that only uses pitch, loudness, and length, yet none the less communicates all the richness of normal human speech. Speech is the use of the
modern vocal apparatus to produce modern human speech sounds.

Perhaps early hominids whistled or hummed to one another like the Pirahãs? Or perhaps they used 'tonally distinguished grunts' like vervet or baboon calls? We do not know exactly how language began. But whatever our ancestors employed as their medium of communication, speech and the vocal apparatus were generations in the future, as were the many effects on the human brain and cognitive evolution that the development of speech produced.

From our imaginary vantage point of the prehistoric past, we can consider the process of the evolution of communication in more detail. First, we need to acknowledge the simple fact that communication only arises with entities that desire to communicate in the first place. Talking about communication before asking *why* people communicate, though, is to put the cart before the horse. We need to think about the unique desire for communication that modern humans share with our evolutionary ancestors to better appreciate the actual mechanics of such interaction. To many researchers, the fundamental question in the evolution and nature of human language is: Why do people want to communicate with each other in the first place? It seems that no other primate feels the need to communicate in the same way that humans do. How does our interest in interacting with our fellow humans differ from the interest that other apes demonstrate for their companions? The principal difference is the complexity of our relationship to other humans – our beliefs, knowledge, and values (effable or ineffable, that is, spoken and unspoken). These can all be corraled together into the term culture. And it is culture that works together with our desire for interaction and our mental and physiological limitations to produce language.

An understanding of human language and communication presupposes an understanding of culture. To some, Culture (written with a capital 'C') would be the refined accomplishments of a civilization, things like visual arts, music, architecture, literature, and so on. That is one way to look at culture. But it is not what I am interested in here. Culture is something that each and every human possesses, not merely the Mozarts or the Miles Davises. Yet the problem we run into when we attempt to offer a single, all-encompassing definition of culture is that, like language, culture is an abstraction. If you do not believe me, think of a question like, 'What is American culture?' This has no definitive answer. I am not denying that there are broad similarities between, say,

Californians and New Yorkers that could be called cultural. I am saying that, just as clearly, there are significant differences between inhabitants of these states, differences ranging from the ways that they dress to the things that they eat to the entities that they worship, as well as their behavior and views on a variety of topics. As a fourth-generation Californian, I can say also that within the state of California the differences between one region and another, or between one economic class, one age group, one educational division, etc. and another, are no less profound than the inter-state cultural contrasts between New York City and Los Angeles. Anthropologists used to think that they had a pretty good idea of what culture was. One of the most widely cited definitions of culture comes from the early British anthropologist Edward Tylor, who was working at the end of the nineteenth and beginning of the twentieth centuries:

> *Culture, or civilization, taken in its broad, ethnographic sense, is that complex whole which includes knowledge, belief, art, morals, law, custom, and any other capabilities and habits acquired by man as a member of society.*

There are two problems with Tylor's definition of culture, however. On the one hand, it does not really tell us how the distinctiveness of one culture could be described. What makes Hopi culture different from Ainu culture, for instance? On the other hand, it fails to capture the variability between cultures in any clear way.

Culture is hard to define. Over the years, and especially in modern anthropology, most students of our species have abandoned the effort to draw sharp boundaries around an idea of 'culture'. Rather than say that people 'have' a culture, many anthropologists would instead say that people 'live culturally'. This is an important idea, but it still needs to be fleshed out with a better idea of what a culture could be. A working definition is that living culturally is sharing a similar set of ideas and values.

For example, you and I share several ideas, such as the idea that information can be put into writing. And we share a value – that of increasing our mental store of new information – otherwise you would not be reading this book. But not all people share this value or this idea. We may well share the value that excess body weight is bad (even more basic is our shared belief that there *is* such a thing as 'excess body weight' in the first place). We probably also share the value that stealing is wrong and the very idea that private property exists – the value that underlies the notion that stealing is wrong. Just based on these values, you and I live in overlapping cultural ways, guided by, in these cases, some (but never all) of the same values and ideas. Or, as some would still want to put it, we share a culture.

At the same time, we also differ in profound ways. You certainly have values that I do *not* share. Maybe you have adopted the values of veganism whereas I might be a hardcore carnivore. If so, then we live in culturally similar ways with respect to some values, but differently with respect to other values.

What I am saying is that 'culture' is an abstract notion that applies to shared sets of values and ideas, recognizing that no two people share all the same values. The more ideas and values that two people share, the closer they are culturally, the more alike their 'cultural living' is. That is why talk of 'having a culture' is misleading. The more values and ideas we share, the closer to each other we live culturally.

Culture is the field in which the mind grows and creates – the field fenced in by shared ideas and values. Constrained by these ideas and values, members of the culture create the perimeters of their existence and their means of survival. Could language as we know it just be one of the many creations of culture, a tool of human invention?

For all of us, knowing something about the nature and origin of human language and the debates that surround these issues can help us to understand ourselves and our fellow *Homo sapiens*. There are two broad questions of interest. First, *why* we do talk to one another and how did evolution prepare us for this? Second, *how* do we talk to one another? In other words, how are our interactions structured? To answer these questions we would need to figure out how evolution changed our species over time in order for us to formulate and transmit specific, discrete meanings.

Looking more closely at the process of thinking itself, and the myriad functions of human reasoning, it is immediately obvious that the most important thinking tool at our disposal, besides our brain, is not a calculator, a book, or a computer, but language. Thinking is possible without language. My dog does it. But non-linguistic thinking does not get us very far. Without language most concepts would be ineffable and unthinkable. No math. No technology. No poetry. Minimal transmission of thoughts from one mind to another. And without language it would be impossible to sequence our thoughts well, to review them in our minds, to engage in contemplation.

At the same time, language is not an unmixed blessing. It creates some problems. Arguments, declarations of war (not war itself), lying of any complication, and so on would be very difficult – impossible, in fact – without language. Only animals with language can engage in these activities. And it is far from perfect. The languages we speak are full of defects. Many of the divorces in the world are at least partially caused by poor communication – misunderstandings abound between people because their feelings and thoughts are so often inadequately communicated, even in principle, by the languages that they speak. Part of this is cultural, part gender-difference-based, and a lot is due to the vagueness and ambiguity inherent in language. This is further evidence that language evolved because evolution never creates perfection, only adaptations that are better than those which came before and are 'good enough' to help us survive longer than our contemporaries.

Language helped us become human. There are few other tools that we can say that about. Communication made it possible to further develop and share our values. And shared values enriched our ties of culture.

Yet I and my fellow researchers are still searching for the answers to some of the biggest questions about language. We cannot even agree about what it is for. Did it evolve for communication or for the better expression of thought in individual brains? We have not yet begun to plumb the profound diversity between languages. We know so little still about language that the best we can do is to develop a set of alternative views to provide intellectual themes for ongoing testing and research.

The view of language as a cultural tool, rather than a biological one, is one such alternative. But this raises the important question, if language is a tool, what is it a tool for? It has two principal functions: thinking and communicating. Let us look at both of these, beginning with language's role as a tool for communicating. In this role language does a number of things, the most impressive of which is that, in some circumstances, it can immediately change the world.

The Oxford philosopher J. L. Austin put it best in the title of a short book that became one of the most influential treatises in the history of language studies, *How to Do Things with Words* (1962). How *do* you do things with words? Austin was by no means the only person to recognize the power of language as a tool. This importance has been recognized in most societies and in most subcultures.

The Bee Gees, as the rest of us do implicitly, recognized the power of language in their 1968 lyric 'It's only words, and words are all I have, to take your heart away.' The power of words is not only a thread running through popular culture, it is also a common theme in serious western literature. Perhaps the best example of this is *Cyrano de Bergerac*, the 1897 play written by Edmond Rostand, in which Cyrano writes love letters for his companion Christian de Neuvillette to help him win the heart of the beautiful Roxane. Christian is handsome and wealthy, but he lacks skill with words. Cyrano comes to his aid, writing beautiful prose and poetry to Roxane, under Christian's name. But Christian's plan backfires – Roxane falls in love with the author of the words that have so moved her, Cyrano.

Bible translation, to take another example, makes no sense unless translators believe in the power of words (God's words in particular). Words change the world. They do this because we let them do it. Because we want them to do it.

This power is so great that it gives rise to expressions like 'The pen is mightier than the sword.' The power of the pen comes from meaning, sounds, and grammar – the bones, blood, and flesh of language.

We employ words to arouse love, hate, fear, affection – the full gamut of human emotions. Words and the grammar that glues them together are essential to the expression of our thoughts to ourselves and to one another. But language is so intimately part of us that sometimes we find it hard to see how complicated it can be without some examples.

One example that caught my eye was the 2009 trial of pop music innovator Harvey Phillip ('Phil') Spector. Spector became famous for his impact on Top 40 music through his songwriting, his guitar playing, and, especially, his producing. His career took off in 1958 after a song he wrote charted at Number 1 in the US, the Teddy Bears' recording of 'To Know Him is to Love Him.' He followed this up with other songs, such as the 1961 smash success 'Spanish Harlem,' co-written with Ben E. King. Not only that but he was a popular session musician (it is Spector's guitar solo we hear on the Drifters' 1964 hit 'Under the Boardwalk'). From the early 1960s Phil Spector produced albums for the Beatles and many other artists, increasing the orchestration and number of instruments over what was common in recordings of the era, inventing what became known as 'The Wall of Sound.' There is no doubt that Phil Spector affected the tastes and memories of many pop music lovers.

But beyond the creative lyrics and music we have all danced to and hummed along with, there apparently lurked a troubled, shadow-darkened soul. After playing Russian roulette over the years with several attractive women he had picked up, he finally killed one of them – the lovely Lana Clarkson – on February 3, 2003. For that death, on May 29, 2009, Los Angeles Superior Court judge Larry Paul Fidler looked at the 69-year-old Spector and pronounced: 'I hereby sentence you to a minimum of nineteen years in prison.' From the moment those words were uttered, law enforcement officers were entitled to restrict Spector's movement and personal liberty until he turned eighty-eight years of age. Before those words were spoken, no one knew whether or how long to keep Spector away from the rest of society. Afterwards, his eighty-eighth birthday seemed far away on a very distant horizon. If this seems obvious, it is because we take the social power of language for granted.

Words change the world. The words I have chosen above to illustrate this story belong to a special class of verbs which include 'to sentence' and 'to pronounce,' and also 'to arrest,' 'to promise,' and such. Philosophers and linguists refer to these more technically as 'performatives' – words whose very utterance performs an act. Such words have the effect they do because societies endow them with a certain power. All of these cases illustrate the cultural foundation for the words that made the changes. J. L. Austin was the first to discuss the importance of such words and how they get their power. His student, the philosopher John Searle, has made the use of words to accomplish things, what he calls 'speech acts,' a part of every intellectual's understanding of human language.

Let us think again about Phil Spector's court case in light of speech acts. If the judge had said something like, 'Buddy, you need to spend the rest of your life behind bars,' or even, 'Mr Spector, I want to ask you to go to jail now. We'll let you out in nineteen years,' the bailiffs would have made no move. Those words are not the words that are authorized to have the effect of depriving someone of their liberty in a US courtroom.

At the same time, although the use of the appropriate words by the judge was important, this did not alone settle Spector's fate. First, the judge needed to be vested by his society with the authority to utter them with legal power. Second, even this person has to pronounce these words in the right context for the words to be effectual. Assume someone meets a judge outside their courtroom for dinner. If the judge were to look over his steak at dinner and say to that person, 'I hereby sentence you to life in prison,' his or her words would be interpreted as an attempt at humor. This is because we all know implicitly not only that a judge must have the legal authority to utter certain words, but also that his or her authority requires a particular setting and a particular sequence of events. The judge must speak in the appropriate legal location and after a trial that follows a prescribed format – defense and prosecution statements being made by official defense and prosecution attorneys, perhaps the presence of a jury, and so on. Once these conditions are met, then when the words of sentencing are spoken they are taken deadly seriously. All of these conditions were met for Mr Spector. For him the judge's words transformed his life and took away his freedom. They were no dinner joke.

In a marriage ceremony, just like in a murder trial, conditions must be met before words can perform actions. In the case of a marriage these conditions include the requirement that a minister be ordained by a legally recognized church or that the person officiating possesses an alternative form of legal authority. Next, the couple must intend to be married. That is what saves people who are married in the movies from being married in real life – movie marriages do not involve the intention to marry. As far as that goes, the minister must also intend that his or her words be the legal declaration of marriage when uttered. This matter of intention is also why the participants need to be in their right mind – for example, if the groom were drugged or the bride senile, their marriage could be annulled. Intention is necessary for the marriage to be real and legitimate. But so are the right words. Without the correct words, the intention and the place are ineffectual.

Arresting someone is just the same at this broad level. In an episode of *The Andy Griffith Show*, aired on December 16, 1963 and entitled 'Citizen's Arrest,' Jim Nabors ('Gomer Pyle') is arrested by Don Knotts ('Barney Fife'). The comedy is based on what happens after Barney gives Gomer a ticket for making an illegal U-turn in his car, greatly irritating Gomer. When Gomer sees Barney immediately commit the same U-turn after ticketing him, he runs after Barney yelling, 'Citizin's ar-RE-uhst, citizin's ar-RE-uhst!' Gomer tells Andy that Barney broke the law. Andy agrees. An indignant Barney goes to jail rather than pay the fine.

Did Jim Nabors place Don Knotts under arrest in real life? Of course not. The arrest on *The Andy Griffith Show* was different from a real life arrest. We cannot explain the difference simply by saying that someone arrested in real life must actually commit a crime but that Don Knotts had not. Being placed under arrest is independent of being guilty of anything. After all, that is why we have trials. Our legal system assumes that every arrest is a legal error – that the police have imprisoned the wrong person ('innocent until proven guilty'). The television arrest of Knotts was not an arrest in real life because the intentions and authority were lacking, even if the words were right. If you hear the words 'You are under arrest' directed towards yourself in the right circumstances, you are under arrest whether you broke the law or not.

Philosophers call these prerequisites for enabling words to become society-changing actions the speech act's 'conditions of satisfaction.' A speech act is any act in which one makes a declaration, asks a question, makes a promise, or so on – the act of one human speaking to another in order to change their knowledge or behavior or to otherwise affect them. Speech acts are fundamentally cultural – although all cultures distinguish questions from statements, not all cultures make explicit promises or condemn or declare war. But in any culture, the range of speech acts that are used are governed by general principles by which all members of the culture recognize that a speech act of a certain kind has taken place. These principles are the act's conditions of satisfaction.

It is amazing that the evolution of our species and its cultures has resulted in word power – that the sounds emitted from our mouths can change the world, marrying, arresting, condemning, promising, christening, and so on.

Thus communication is not always a simple exchange of information. Language is a complicated and powerful tool. It is not a small matter to learn to use it. And this is true in all cultures, in all languages, even though the way to use a language in one culture may not be the best or even an appropriate way to use it in another.

Evolution provided a reasonable solution to human communication in the form of human language. As I said, imperfections of language are the source of many spousal and professional disagreements, confusions in the courtroom and classroom, different understandings of directions given (a map's visualization of spatial relations is much more effective at helping people locate themselves than directions given in spoken language), and so on. On the other hand, these imperfections are to be expected in any development programmed by evolution. They should not be taken as evidence that the function of language is something other than communication, a claim that is made from time to time.

Over a period of time, an evolutionary solution will improve or disappear as a creature's environment favors or disfavors it. A solution will be just as good as it needs to be to ensure survival of the most offspring. No better, no worse, in most cases. The term suggested by Nobel prize-winning economist Herbert Simon for this characteristic, whether it is found in biology or economics, is 'satisficing.'

The concept of satisficing is important to our understanding of how language can be so widespread in spite of its defects, like ambiguity and vagueness. For now, we can simply say that evolution gave us language. But what exactly does talking in the sense of 'orally channeled communication' entail? How did we need to evolve to get us to our current levels of language ability? If we think of language as a tool to solve a problem, then the first step towards understanding language is to understand the problem it must solve, communication.

In the late 1940s, a group of scientists and mathematicians turned their considerable skills and intellects to address this very issue – the nature of the problem. Their motivation, since most of them worked for the company that was then called Bell Telephone, was to build a better telephone system. Though it is often difficult, and usually pointless, to try to single out one person as the *primus inter pares* in endeavors of such complexity, in the development of communication theory one name is especially prominent – Claude Elwood Shannon. Most subsequent communication theorists recognize Shannon as the Captain Kirk of communication theory, the original trailblazer whose work was built upon by those who followed. Although Shannon's work is highly technical and difficult for the average person to read, we are fortunate that he had an interpreter, Warren Weaver. Weaver's introduction to Shannon's ideas in *The Mathematical Theory of Communication* (1949) was influential, bringing these ideas to a much wider audience than would have been able to read Shannon's original, mathematically challenging contributions.

As Weaver and Shannon laid it out, communication is a multi-step process and its multiple components must be analyzed individually. The same holds whether we are talking about the research of Bell Telephone company engineers trying to design the world's best telephone system or evolution producing language in humans. Evolution had to prepare early hominids to solve this problem if they were to communicate with one another and especially if language, the ultimate communicative tool, were to appear on the scene. If the outcomes of evolution failed to include a solution to all of these steps, communication would not be possible.

Imagine a typical act of communication. There is an individual with a message. This message is some unit of information such as a word, a story, a sentence. There is a person or people that the speaker intends to pass his or her information along to. That is simple enough. But Shannon made it all clearer in a diagram (see Figure 1).

Getting a message across results in a shrinking of the world. Whenever anyone says something, they focus their audience's attention on *that*, whatever it is, and away from everything else around them. They narrow the universe of information and facilitate both thought and

## Figure 1  A Model of Communication

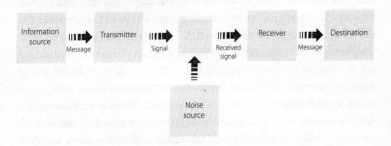

communication. This understanding of the notion of 'information' is a technical concept. It is easily understood, but we must be careful to shed our common understanding of what information is to appreciate this important technical meaning. We can understand information in the technical sense if we imagine all the things there are to talk about. It is perhaps infinite. But culture and circumstance will severely restrict what might be said in a particular exchange between two people. And as most of us living in the twenty-first century know, information in this formal sense, the sense of the computer age, is measured in bytes. This involves logarithms and some other tedious concepts that fortunately do not need to be discussed here. But it is useful to appreciate that human cultures narrow the things there are to talk about and keep the flow of information manageable.

Since it is vital to an understanding of communication and how this shapes the form of our language, I will take the diagram above apart and look at each bit of the problem more carefully. In doing this, I will ask you to imagine that you are giving a public lecture and that you want your lecture to be clear and understandable to your audience.

This diagram can be translated simply enough. The 'information source' can be one of many things, including a human speaker, a computer or a radio, etc. In pre-human times, the source would have been a small ape.

This source selects one message out of a set of possible messages. So I might choose to say, 'Come here,' as opposed to, 'Go away' or, 'Let's eat,' etc.

In the specific technological application that Shannon and others had in mind, a landline phone, the channel is a wire and the signal is a varying electrical current. When we speak face to face, the channel is the air and the signal is varying sound pressure. With cellular phones, the channel is space and the signal is an electromagnetic wave. The same communicative principles hold across different channels and signals – part of the brilliance of Shannon's insights.

Looking at Figure 1 in more detail, we see that the first thing needed for communication is a source for all the information you will be passing along to your audience. This source of information would be you. In fact you are the source of information in this particular act of communication even if you are quoting someone else or plagiarizing. Shannon's diagram is not concerned about 'ultimate sources' but with the local, immediate source in a particular act of communication. The information is coming out of *your* mouth and it has presumably passed through your brain in some form first. So somewhere inside of you – hopefully your brain – will be the ultimate source of information in this communication flow.

Alright. But since there is no mental telepathy in our non-science-fiction world, you need a way to get the information out of your brain and on its way to your audience. This is easy. Your mouth, your hands and the rest of your body serve as one complex transmitter for your brain. No healthy individual entrusts the act of communication solely to their mouths or their hands. We all engage our hands, mouths, facial expressions, body orientation and body movements whenever we transmit any sort of message from our brains towards an audience.

To recap: our brains are the source of information as we communicate and the transmitter we use is our entire body, although most of the burden normally falls on our vocal apparatus – the set of organs we use to generate and amplify speech sounds, including our tongue, teeth, vocal cords, lips, lungs, mouth and nose. The human vocal apparatus shapes and emits acoustic energy. Humans make sound waves. If you like, the lungs are like a balloon and the human head and neck are like the tip of the balloon. As a child, I used to enjoy squeezing the tip of a full balloon and make whiney, screechy noises as the air came out. Our vocal apparatus does exactly that – it stretches and changes form to make different sounds from the air coming out of our lungs. Modern humans need to make a large array of sounds in this way in order to produce acoustic energy rich enough to convey the range of our thoughts in an act of communication. Although our ancestors may have gotten by fine with grunts, it isn't easy to use grunts (though it would be technically possible) to provide a financial analysis of the British economy. Words of all sorts make this task easier. And a way to get the words from me to you – some energy traveling across a medium like air, water or light.

While it is technically correct to say that all we produce when we speak is acoustic energy, there is clearly more to it than that. This acoustic energy must carry a meaning capable of interpretation. What comes out of our mouths is not just sound waves. The real products of our transmission are things like slogans, jokes, sentences, stories, poems, words, and many combinations thereof. Each item on this list has two distinct and vital facets: meaning and form.

Now, where does this signal of meaning and focus go before it gets to your audience? The physical properties of your signal take the shape of sound waves traveling through the air. The energy burst that proceeds from your mouth through the atmosphere to the ears of your audience has complex physical properties that human brains are equipped to decode. In order for them to understand what you said, the brains of your audience must be able to reconvert the acoustic energy in the atmosphere into the forms of signs and then associate these forms with the correct meanings of the signs.

How do hearers do this decoding? They rely on their biological and cultural platforms. Later on I am going to discuss in more detail how ears do what they do, how the process of speech perception works in a (very small) nutshell. Roughly, the acoustic energy is transformed by the ear to electrochemical energy that is passed from the ear to the brain. Sometimes the brain is unsuccessful. So the hearer might acknowledge this by asking the speaker, 'What did you say?' Amazingly, though, most of the time the process seems to work. This is a major evolutionary accomplishment.

In a normal utterance, the brain first has to recognize words. But words alone are insufficient to convey a message worth giving to a formal audience. No audience will assemble to hear someone recite the phone book. Even if they fill the hour with their recitation, that is not what the audience came for.

But there are a couple of ingredients of communication lacking from the model above. These are syntax and coherence. Syntax refers to the order in which words are presented and to the way that they are grouped with one another. This can be complicated in some languages, but is often quite simple. In their 1985 book *Generalized Phrase Structure Grammar*, linguists Gerald Gazdar, Ewan Klein, Geoffrey Pullum, and Ivan Sag showed that a large portion of English grammar, often thought of as very complex, could be summarized in a surprisingly short list of rules.

There is more to getting a message across than syntax in any case. One ingredient of communication that no culture can do away with, because it is such a vital part of the communication glue, is what many linguists call 'coherence.' Coherence is the property that groups of words and sentences are said to have when they 'make sense.' Coherence is especially important for putting sentences together to write a book, tell a story, or create any message exceeding one sentence in length, such as the imaginary speech you are about to give.

If your message consists of the three sentences below, you will probably not be understood well:

*Britain has now had more than a decade of continuous Labour rule.*
*Carol's pumpkin is the cause of the widespread fright in October.*
*Billy Bob never gave Hugh Grant a chance when they were on screen together.*

Each of these sentences is fine on its own. We can understand them individually without difficulty. But put them together and they become confusing. They do not work together. They do not contribute to a single story. Perhaps some context could be given to make them fit with each other better than they do now. But as they stand they do not cohere.

Other sets of sentences do cohere.

*Jack and Jill ran up a hill to fetch a pail of water.*

*They took longer than expected, though.*
*I wonder what they did up there?*

These sentences can be understood as all about the same thing, the questionable hill endeavor of Jack and Jill. Each sentence of this story contributes something to the understanding of the event. And they are woven together by their grammar. For example, the use of the personal pronoun 'they' in the second and third sentences refers back to Jack and Jill and establishes a dependence between the sentences. In the third sentence, the pronoun 'there' refers to the top of the hill we are talking about, again linking the sentences. That is one of the functions of pronouns, they let us know that the sentence with the pronoun in is part of the same story. Proper names like 'Jack' and 'Jill' are used at the beginnings of stories usually. (Rarely do you begin to tell a story with a pronoun.) But then they are abandoned in favor of pronouns. Think of the effect of reversing the pronouns and proper names in the Jack and Jill story:

*They ran up there to fetch a pail of water.*
*Jack and Jill took longer than expected, though.*
*I wonder what Jack and Jill did on the top of the hill?*

Suddenly the story no longer coheres. The hearer is not sure who 'they' refers to even with 'Jack' and 'Jill' coming in the next sentence. Coherence is a matter of culture and style, not rigid syntax, though. In the Pirahã language, pronouns are used differently (they are used less than proper names, the reverse of English and many other languages) and the story just given would be like this:

*Jack and Jill ran up the hill to fetch a pail of water.*
*Jack and Jill took longer than expected, though.*
*I wonder what Jack and Jill did on the top of the hill?*

Humans need coherence. I am not talking merely about finding meaning in our lives, ensuring that our pasts, presents, and futures cohere. But in the stories we tell. We cannot figure out how to decode the signal coming into our ears without being able to assume and look for clues to the coherence of the message we are hearing. Being able to make stories cohere is something that varies among members of a culture, though. Not everyone is equally good at it. That is why some writers are said to write more clearly than others.

In the 2000 movie *Memento*, the main character, played by Guy Pearce, loses his short-term memory due to a severe head trauma. But he partially (and this is where the suspense and mystery of the movie come in) compensates for his loss of memory by keeping copious notes. The most important notes, things he must never lose sight of, he tattoos on to his body.

Even though language, as the movie shows, is not an adequate replacement for a good short-term memory, it is a vital component of the way that we remember and can even enable us to 'remember' things we forgot we knew, or indeed never knew. History is the ultimate example. By writing down their stories, or memorizing them carefully and transmitting them as oral literature, most societies are able to build a 'collective memory' of their past. No other species can actively do this. No collective memory of the human type is possible without language. It is the greatest of all our mental tools. The greatest of all our cultural tools. This is uncontroversial. What does provoke discord, however, is the source of this tool, not only in the species, but also in the development of individual humans.

The source of knowledge of all kinds in individual humans has long plagued philosophers.

*Chapter Four*

# DOES PLATO HAVE A PROBLEM?

'*The world does not speak. Only we do. The world can, once we have programmed ourselves with a language, cause us to hold beliefs. But it cannot propose a language for us to speak. Only other human beings can do that.*'

Richard Rorty, *Contingency, Irony, and Solidarity* (1989)

The origin of human knowledge is one of the oldest and most controversial issues in philosophy. Since Chomsky's work from the late 1950s to the present, in which he argues for the existence of an innate universal grammar, the origin of knowledge has also become a central issue of modern linguistics. The main question is how evolution has prepared humans to acquire languages. Did natural selection or some other force build a special knowledge of language into all *Homo sapiens*? Or does our use of language stem from the non-language-specific abilities of our species, such as our intelligence?

The hypothesis that human beings have been 'groomed for language' by evolution is supported in numerous ways. Some researchers argue that children do not learn language like they learn other tasks. Rather, children acquire language with a rapidity that can only be explained by the fact that they are hard-wired for language at birth. Other researchers claim that this idea is supported by human biology and that there is evidence for a 'language organ' and specific genes and regions or tissues or neuronal networks of the brain that are specialized for language and found congenitally in every member of our species.

We all know that language exists, that it is useful, and that every normal human uses it. How it came to exist and the ways in which it has taken on usefulness are still unresolved problems requiring solutions from modern science. As I have already shown, one solution for the problem of how language exists is to understand language as a cultural invention created to solve the communication problem. But the other principal alternative is to conceive of language as so rooted in human biology that all humans are born with specific knowledge about language structure. The next step, then, is to try to find out whether there is any convincing evidence that specific innate knowledge or specifically linguistic biology in humans exists. Like so many important philosophical issues of our civilization, this one was first placed on our intellectual agenda by Plato, the fifth-century BC philosopher who studied with Socrates and whose most famous student was Aristotle. If we begin by looking at the idea that people are born knowing things about language, we can move from that to the question of whether components of human biology are specifically linguistic.

Everyone who believes in innate knowledge of the kind that some linguists, philosophers, and psychologists claim for human language owes an intellectual debt to Plato. Even the Bible owes a nod to Plato. According to some traditions of biblical scholarship, the gospel of John was written by Jesus's youngest original disciple, John, a poorly educated fisherman (the gospel does seem to be written by a non-native speaker of Greek, due to its small vocabulary and simple syntax). The gospel begins with a synthesis of Christian and Platonic beliefs about truth and origins (my translation from the Greek):

> In the beginning was the word ('logos') and the word was with God and the word was a god/God ...Through the word all things were created. And the word became flesh and lived with us.

Plato uses the same Greek word, 'logos,' to mean 'ultimate essence,' the underlying truth of all things. This word was selected because in Platonic philosophy the true meanings of words reveal the essences of the things that the words name. This Platonic concept, which predates the New Testament by about five hundred years, is responsible for the modern convention of naming fields of study by adding a derivative form of the Greek word 'logos' to the end of the word representing the matter under study. Thus the study of humans is 'man' ('anthropos' in

Greek) plus 'logos,' or 'anthropology.' This would literally mean 'the ultimate truth or essence of man.' For the study of life (in language and in biology) we get 'zoon' 'animal' plus 'logos' zoology. For the study of the stars we get 'astron' 'star' plus 'logos' astrology (later, because of non-scientific applications of this term, the term astronomy was introduced, from 'astron' plus 'nomos' 'law'). And so on.

Platonic philosophy has exercised a greater effect on modern science than the mere invention of names, however. The 'logos' at the ends of these words is an explicit claim that we can, with sufficient diligence, luck, and intelligence, know things truly.

John agrees in his writing that there is an ultimate, divine truth. That is why he used Plato's term 'logos.' But in the last sentence of this quotation, John goes beyond anything that Plato ever claimed. John says that this divine 'logos' came to earth to live in the form of Jesus. John reverses Plato's view of truth. Plato says that we cannot directly apprehend the truth with our senses. We perceive only shadows here on earth. We are separated from heaven, where Plato claims that truth is found. According to Plato, philosophers can come closer than any others to the understanding of the heavenly truths. Yet John, the fisherman disciple, says that all people can apprehend truth in this terrestrial life – in Jesus. John's innovation paradoxically moves us closer to empirical science through the non-scientific notion of the incarnation of God in the form of Jesus, which challenges us to know truth now in the world around us.

Plato's influence on Christianity is seen in another way, however. There is a vital point raised by his heavenly view of truth that has exercised thinkers for centuries. Plato observed that the source of some human knowledge is hard to identify. People seem to know a lot more than they could have learned. Plato claimed that this was so because everything, including our souls, is found in a true form in heaven, transcending the time and space of our earthly lives. Thus, he reasoned, our souls would have learned everything in heaven, prior to living on earth. The purpose of the philosopher's inquiries into truth in this world is therefore to *recall* the truth we already know. We learn nothing. We remember all.

Plato makes this point, as he often does, by constructing a dialog between Socrates and a poor soul who realizes by the end of his exchange that he has bitten off more than he can chew by disputing with the master philosopher. The particular passage of Plato that goes to the heart of the issue of learning *v*. recall takes place between Socrates and a friend, Meno. Although the Meno dialog is mainly about virtue, it is a brilliant defense of what is known as a priori knowledge, that is, knowledge gained before experience. Plato stages the conversation at a dinner at Meno's house, where Meno puts the following question to the great man: 'Can you tell me, Socrates, whether virtue is acquired by teaching or by practice; or if neither by teaching nor practice, then whether it comes to man by nature, or in what other way?'

The answer given is that virtue is indeed not learned. This puzzles Meno, who then asks, 'But what do you mean by saying that we do not learn, and that what we call learning is only a process of recollection?'

Socrates answers by calling to himself one of Meno's servant boys and leading him through a series of questions about geometry, in order to establish that this uninstructed child possesses surprising and accurate knowledge of the Pythagorean theorem.* The philosophical question that is at stake here for Plato is: How can people know so much about things that they have neither been instructed in nor have had any direct experience of? Since Plato, the proposal that a priori knowledge exists has had a long and controversial history in western thought. All theories of the nature and origin of human language fall on one side or the other the debate begun by Plato.

The centrality of a priori knowledge in discussions of language stems mainly from the work of Chomsky, who pioneered the attempt to link a scientific theory of human language to the nature of the human mind. Chomsky argued that language learners are in the position of Meno's slave: they know more about language than they seemingly could have learned. Chomsky calls the question of how we come by such knowledge 'Plato's Problem,' in deference to Plato's pioneering thought on a priori knowledge. Subsequently this phrase has been used to refer to situations in any other area of human cognition in which there is evidence that people know more than they could have learned.

Since researchers have claimed that there are many such phenomena,

---

* For those interested in reading the full dialog, see Appendix p. 328.

Chomsky is far from the only voice to support Plato's and Socrates'(at least insofar as Plato accurately represented Socrates) view of knowledge. Many others have explored the power of a priori knowledge in explanations of math, music, morality, and other domains of human cognition (reporter Nicholas Wade of *The New York Times* writes regularly about these proposals). One of the most prominent researchers on a priori knowledge is Harvard psychologist Elizabeth Spelke. Her work is among the most interesting and thought-provoking research in the world on Plato's Problem. Steven Pinker, another Harvard psychologist, and Marc Hauser, a biologist, are also well known for their work on things like the 'language instinct' and the innateness of human morality, respectively. Perhaps the dean of all innateness explanations of human behavior and thinking, however, is E.O. Wilson, whose 1975 book *Sociobiology* became the basis for an entire school of thought, known as evolutionary psychology.*

Chomsky's advocacy of Plato's idea of an a priori knowledge, especially the view that the core properties of human language are innate, has been one of the most influential proposals in modern intellectual history (but for Chomsky such knowledge presumably originates in our genes, rather than in heaven). Professionals from a gamut of scientific fields of study are fascinated by the idea that we know language before we are born.

At first blush, this idea feels right to a lot of people. A good deal of evidence suggests that children do learn language with amazing speed and that they know things that are equally astounding relative to their ages and experience. Several bits of suggestive data have also been used to support the existence of Plato's Problem and Chomsky's nativist solution to it ('nativism' is the idea that certain knowledge or characteristics are part of our genetic endowment; it is a neutral term). First, some genes, like FOXP2, seem to be specialized for language. Second, children seem to learn language in well-defined stages no matter what language they are learning or where they are in the world. Third, Chomsky

---

* Not everyone would agree with my claim that *Sociobiology* was the intellectual sire of evolutionary psychology. But the conceptual connections are not hard to see.

claims that all children learn their languages with equal speed and accuracy, accomplishing all of this before puberty.

A fourth claim made in support of the nativist view is that there is a critical period, roughly running from birth to puberty, after which our ability to acquire new languages seems to change dramatically, making the learning of subsequent languages much more difficult and using different cognitive abilities than are used in first-language acquisition. The problem with this is that in order to establish clearly the existence of a critical period one would have to show that second-language learners after puberty never reach the same level of fluency or control of the language as first-language learners. Next, assuming that one did establish this language attainment differential convincingly, it would be necessary to show that the second-language learners gave as much time and effort and were exposed to as much data as first-language learners.

Elissa Newport, Jacqueline Johns, and many other researchers have published papers in support of the critical period hypothesis, papers that are widely accepted in the community of researchers who believe that language is innate, as proof that such a period exists.

Nevertheless, there are alternatives, both empirical and conceptual, to the notion of a critical period in language acquisition. Conceptually, the French psychologist Jean Piaget, the founder of behaviorism, B.F. Skinner, and other major psychologists have proposed alternative explanations of the data. Empirically, there is still a lot of work to be done to show that there is evidence that all children learn language equally well and that those learning second languages all fail to achieve native fluency under the same conditions in which children acquired their first languages. And even if this differential accomplishment between adults and children under identical conditions can be shown, it is not clear that we need to appeal to a specifically linguistic critical period to account for these facts.

Piaget, for example, has written extensively that, even if children learn languages better than adults, this would likely follow from the maturation of their brains coupled with the fact that language learning is part of the general task of building a distinct identity in the world. Therefore, since language learning is a different task culturally for children than it is for adults, this could explain any differential facility in the acquisition of languages without proposing a universal grammar or other language-specific cognitive device. Piaget does believe that learning is innate in a sense, but not that there is anything language-specific in our genetic endowment.

There are simply no uncontroversial results establishing either of these claims. Therefore, the idea that there is a special period in which an innate language acquisition device is functioning in normal humans, while interesting, is far from established truth. Other hypotheses are still free to explore this space.

It can also be appealing to believe that knowledge of language is built very specifically into human biology because there is a widespread belief that some parts of the brain are specialized for language. If this were the case, then we might expect that the outputs of these special-ized portions of the brain, human languages, would share most of their essential features, regardless of where they are found in the world. To put it in Chomsky's words (see http://bigthink.com/ideas/16057):

> ... the apparent diversity [of human languages] is pretty superficial. So that if, say, a Martian was looking at humans the way we look at, say, frogs, the Martian might conclude that there's fundamen-tally one language with minor deviations. And I think we're moving towards an understanding of how that might be the case and it is pretty clear that it has to be the case. The time of development is much too shallow for fundamental changes to have taken place and we know of no fundamental changes.

It has also been claimed that language is so complex, and that there is so little data available to the child, that it cannot be learned. This is just Plato's Problem restated. But no matter how fast and heavy the analy-ses fly, we must not forget that the nature of language is an empirical problem. Anyone denying or supporting Plato's Problem has to come to the discussion with evidence.

Since we are after scientific evidence and not first impressions, though, we need to ask how well the case for the innateness of language has been made, whether in popular books and reports or via more technical articles in scientific journals. One thing that immediately strikes the observer in surveying this literature is that the arguments for linguistic nativism all share a peculiar form. The experimenter shows that the child knows something. If the experiments are well designed, the experimenter then rules out a few ways in which that knowledge might have been learned. The experimenter then concludes that since s/he found no way in which the knowledge might have been learned, the knowledge is a priori, Q.E.D. But this conclusion fails as a proof.

The burden of proof on nativists is higher than merely showing that children possess surprising knowledge and that a couple of ways of learning that knowledge are not plausible. The nativist would also have to show that there is *no plausible way* children could learn that knowledge. But few efforts have been made to meet this higher standard. So it might be worth considering what other kinds of evidence might support the idea that there is a 'language organ' or 'universal grammar.' Because this is so important, I want to review some of the major supports for a language organ often described in the literature. There are many studies that discuss evidence against all of these points. In other words, these are expectations we would have if language were innate, but they have not yet received uncontroversial support.

The first question that comes to my mind is that if language is innate, should not there be some genetic evidence for this? I would have thought so. Others agree. In an article in *Trends in Cognitive Science*, 2002, Gary Marcus and Simon Fisher state the issue clearly: 'The human capacity for acquiring speech and language must derive, at least in part, from the genome.'

And yet, there does not seem to be any strong evidence for this, or not yet. It has not been established that there are *any* genes specific to language. What we do know from genetic studies so far is that there are genes – the best example, widely discussed in the literature, being FOXP2 – that are important to language. Yet finding a gene that is important to language is not the same as identifying a gene *for* language. For example, my lungs are important to thinking. Without air to my brain I would cease to think. But they are not organs *for* thinking. In the case of FOXP2, what we know is that language is only one of this gene's many applications.

FOXP2 is discussed frequently in the literature about language, however, as though it were *the* language gene. This talk was initially based on a 1994 article by Myrna Gopnik, a (now emeritus) professor at McGill University, published in the *Journal of Neurolinguistics*. Gopnik's article linked FOXP2 to language disorders, in which a mutation of the gene seemed to be implicated in a very specific linguistic disorder found in members of a particular family. This was exciting research. After her research was published, other scientists wondered whether Gopnik had in fact discovered a language gene. This would be striking, concrete evidence for the idea that language is innate. The idea that FOXP2 is the language gene is still a widely held view.

But, again, FOXP2 is not a language gene in the sense that it is specialized for language. Because of its multifunctionality, it would be a mistake to label FOXP2 a language gene. More prudently we should conclude no more than that it is a gene that can affect language, as part of a larger range of physiological functions. Marcus and Fisher investigated this gene in detail, concluding that 'FOXP2 cannot be called "the gene for speech" or "the gene for language." It is just one element of a complex pathway involving multiple genes, and it is too early to tell whether its role within that pathway is special.'

What sorts of problems emerged for the original interpretations of Myra Gopnik's pioneering work on the FOXP2 gene that led people like Marcus and Fisher away from the idea that it was a language gene? First, in the language disorders cases that have been examined, the people with affected language all had other forms of cognitive disorders as well. For example, members of the KE family who served as the basis of Gopnik's study (the identity of the family is confidential and they are known only as 'KE') manifested pathologies such as: difficulties in processing word structure and in understanding some complex structures; inability to make themselves intelligible in many contexts; difficulties in the mobility of their mouths and faces, especially the upper lip, in ways that are not associated with speaking; and, most significantly, a reduced IQ – an average of 86 in the affected members of the family, compared to 104 in the unaffected family members. This range of difficulties is to be expected if, as we now know, FOXP2 is linked to multiple physical and cognitive abilities, rather than to language *per se*. For example, FOXP2 seems to underlie the flexibility or 'plasticity' of neural circuits (lack of plasticity could, as a by-product, cause language problems). FOXP2 is also important for the development of the human lungs and gut. Additionally, FOXP2 is important in the control of the face muscles, including the lips. This latter aspect of FOXP2 is relevant to Gopnik's studies because her subjects also had difficulties in pronouncing language. And this gene is important in the regulation of other genes. Moreover, FOXP2 is not even unique to humans. Homologous versions of FOXP2 exist in mice, songbirds, and other creatures.

Now there *is* evidence that FOXP2 has become more specialized for language over the past 100,000 years. But the adaptations of FOXP2 are themselves likely caused by interaction with and shaping by the language tool and human culture as cultural evolution interacts with biological evolution. Therefore, we must conclude that, as important as FOXP2 is for life, it is not a 'language gene.' This should not come as a surprise. Genetics teaches us more generally that the relationship between an individual human's genes (its genotype) and its appearance, behavior, and any other aspects of the individual's observable nature, the phenotype, is non-linear – that is, for the most part it is not possible to link individual traits of the phenotype to individual genes. This is a very important lesson. It is hard to discover one-to-one relationships between genes and phenotypical traits in humans: one gene can be responsible for more than one observable trait and one observable trait can be the result of multiple genes. Genes are economical, each bringing its own range of roles to play in human development as they adapt to fit the forces of evolution. This adaptation and flexibility is a signature feature of hominid evolution. Thus it is highly unlikely that a simple 'language gene' will ever be discovered. That is just the wrong way to think about things.

Well, if not genes, then perhaps there is evidence for a specific or localized component of the brain or other part of human anatomy to support the idea that our biology includes specific design features for language? And should there be evidence of language in our biology, might we find this in specific parts of our human anatomy? Perhaps. So let us consider that curious mass of saturated fats which occupies the major portion of our cranial cavities, our brains. If there were separate organs or, as philosopher Jerry Fodor would have it, specialized modules of the brain, one might have expected to find some anatomical support for this in, say, regions of the brain specialized for different functions. But the evidence is far from clear on this. It has certainly never been established that any part of the brain is dedicated exclusively to language, although if you take an introductory class in linguistics or psycholinguistics at some universities, you might hear assertions to the effect that there are parts of the brain – Broca's and Wernicke's areas – which are the 'language areas.' The evidence suggests otherwise.

In their 1997 book *Rethinking Innateness*, Jeffrey Elman, Elizabeth Bates, and colleagues underscore why we should not jump to conclusions about the significance of the fact that some kinds of knowledge are found in specific regions of the brain:

> ... *everything humans know and do is served by and represented in the human brain* ... *Our best friend's phone number and our spouse's shoe size must be stored in the brain, and presumably they are stored in non-identical ways, which could* ... *show up someday on someone's future brain-imaging machine* ... *The existence of a correlation between psychological and neural facts says nothing in and of itself about innateness, domain-specificity, or any other contentious division of the epistemological landscape.*

They add that: 'Well-defined regions of the brain may become specialized for a particular function as a result of experience. In other words, learning itself may serve to set up neural systems that are localized *and* domain-specific, but *not* innate.' Therefore, regardless of anyone's findings, or claims about brain specialization for specific tasks, we cannot draw any conclusions about a priori knowledge or universal grammar, say, without first ruling out learning as a source of cortical localization, where specific types of knowledge are found in the brain.

Let me say too that nativists are not committed to the claim that a single region of the brain must be linguistic. Some have made this claim, but it is not crucial to nativism. If someone could show that a region of the brain were dedicated exclusively to language in all humans and was not the result of learning, this would strongly support linguistic nativism. But, as I will show, there is no such support. However,

Figure 2 **Supposed locations of Broca's and Wernicke's areas**

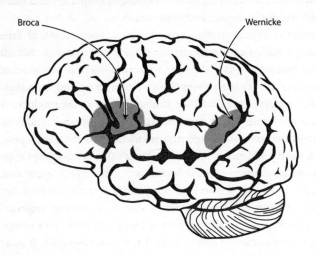

nativists could claim that language is innate but diffuse, its functions spread out through the brain or neuronal networks. This is possible. There is no uncontroversial evidence that brains have or do not have such networks independent of learning. But that is not ruled out in what follows. Rather, what follows is a discussion of the evidence for localized brain specialization for language. This is important because it is part of urban and academic folklore that such regions exist. One of the best known of these regions is 'Broca's area.'

Broca's area is normally identified as a region of the brain located at the *pars opercularis* and *pars triangularis* of the inferior frontal gyrus, that is, on the left side of the brain. Broca's area and Wernicke's area are both indicated in Figure 2.

Broca's area was first identified in the nineteenth century and became associated with the French researcher and physician Pierre Paul Broca. In 1861 he conducted an autopsy on a former patient of his, known simply as 'Tan,' for the reason that the patient could utter no other word but 'tan,' a word that no one else understood. Broca discovered that Tan's brain had damage to the left front portion of his brain, a lesion caused by syphilis. From this discovery Broca drew the conclusion that this part of the brain must be involved in speech (though he was not the first to do this – another Frenchman, neurologist Marc Dax, had made similar discoveries many years prior to Broca).

As a result of Broca's studies, many people came to think of Tan's affected brain region as specialized for language production. But although everyone agrees that there are regions of the brain involved in language, there is also wide agreement that Broca's area is not language-specific. For example, recent MRI findings indicate that the production of speech is not limited to Broca's area. Also, (at least single) lesions to Broca's area alone are not enough to permanently disrupt speech production. A third problem for claims that Broca's area is a language area is that the original lesion studied by Broca in Tan's brain seems not to have fallen in exactly the same area we currently understand to be 'Broca's area.' Fourth, Broca's area can be destroyed leaving language intact. Finally, Broca's area is involved in other kinds of brain activity, such as coordination of motor-related tasks. For example, if you show someone hand shadows of moving animals, Broca's area is activated; the same if someone listens to or performs music. These are not language-specific tasks. Moreover, there are no pure cases of 'Broca's aphasia' – aphasics nearly always have other deficits in addition to language. (Aphasia is the general term for many different types of language and speech disorders, coming from a Greek work meaning 'speechless'.) What such results show is that Broca's area could have a more general function, such as the coordination of activities, of which language production is but one, rather than a dedicated language area.

I do not say that we fully understand Broca's area or that we know with certainty that there are no language-specific regions of the brain. I am saying only that we have not yet discovered such areas. And there's the rub, if you are a nativist. Plato would not have cared one whit if it turned out that there were no regions of the brain specialized for any of the knowledge he considered a priori. This is because, again, for Plato the source of our knowledge was heaven, not the genes. But for Chomsky, evolutionary psychologists, and others searching for the potential sources of a priori linguistic or other knowledge, heaven is not the place to look. The evidence must be in the human brain.

So it would be interesting and not entirely surprising for this view if we found regions of the brain dedicated to language, even though this is not crucial for nativist theories of language. In any case, anyone who says that it has been clearly shown that there are specialized regions of the brain for language which support the idea that language knowledge is specific and hard-wired into us, is wrong. This is false.*

Moreover, new evidence currently being gathered suggests the opposite – that we use different parts of the brain for multiple, diverse functions. Recent findings in research led by Marina Bedny of Rebecca Saxe's laboratory in the Department of Brain and Cognitive Sciences at the Massachusetts Institute of Technology show that the 'visual cortex,' the region of the brain usually associated with vision in sighted individuals, can be used for non-visual tasks, in particular language, in blind subjects, as well as for Braille-reading. This work is extremely important for any attempt to link cognitive functions with specific regions of the brain. And hers is by no means the only research that shows how amazing brain plasticity can be. Whatever is responsible for the localization or specialization of different regions of the brain for different cognition functions, it does not seem to be the result of specific, genetically determined links between functions and cerebral topography. The brain seems no more specialized for language than other individual components of the anatomy are responsible for speech. The lungs, larynx, teeth, tongue, nose, and so on are all vital for non-signed language as the hands are crucial for signed languages, but none of these are either individually or collectively language organs.

---

* A recent series of studies by Evelina Fedorenko and several colleagues at MIT's Department of Brain and Cognitive Sciences offer improvements in neuroscientific methodology. These researchers suggest that previous methodologies for identifying specialized regions of the brain do not seem to work well. According to the newer methodologies Fedorenko and her colleagues have developed, the prospects for identifying brain localization seem more promising. However, even these methods have severe problems. First, they fail to recognize the fact that the portions of the brain that are associated with language are connected to the (non-specialized) basal ganglia. Second, they fail to discuss the fact that chimpanzees and dogs have word-recognition capabilities and, therefore, by their very assumptions ought also to have specialized linguistic regions of their cortex. Third, they neglect to mention that all areas of the brain specialize for tasks that are repeated regularly – they have to be encoded somewhere.

He never says what's good about it. Any positive evidence for anything, Daniel?

The same issues arise for any claim of anatomical specialization for language, such as the other major area of the brain much talked-about as a language-specific region, namely, Wernicke's area, located in the posterior section of the superior temporal gyrus in the dominant cerebral hemisphere of a given individual (for right-handed people this is the left hemisphere; for lefties, it is the right hemisphere). At one time, this region of the posterior temporal lobe of the brain was believed to be specialized for understanding written and spoken language. The evidence for this belief was that lesions in this region, observed originally at autopsy by the German physician Carl Wernicke in 1874, caused a specific type of aphasia. If this sort of claim could hold up, it would be interesting evidence for brain specialization for language, even though it would not in itself necessarily be evidence for the view that language is innate. This is because we already know that brain specialization results from non-innate, cultural learning, as new connections between neurons are formed. Therefore, brain regions that are implicated in language can be important for distinguishing and evaluating various hypotheses on genetic *v.* cultural effects on brain specialization relative to language. So we need to learn as much as we can from such areas, without being impulsive in the conclusions that we draw.

Unfortunately for anyone interested in marshalling anatomical support for the innateness of language, Wernicke's area, like Broca's area, is not exclusively specialized for language. First, just as with Broca's area, there are no agreed-upon definitions of the location and extent of Wernicke's area. There is a general range that people agree on, but no consensus on the exact coordinates of that position in the brain. That makes it difficult to say with any precision that there *is* such an area. Second, recent research shows that this region is connected to other regions of the brain that are, as was the case with Broca's area, far more general in their function than language: e.g., motor control (and pre-motor organization of potential activities). Third, as Saxe's research indicates, even if we did find an area specialized for a particular function in one or a million subjects, the next subject we meet could in many cases be using that area of their brain for something else, depending on their individual developmental history.

Then what are we to make of what is currently known about the association between language functions and Broca's and Wernicke's areas? The lesson is that parts of the brain develop in each individual as the homebase for multiple, though related, tasks, such as language, motor control, sequential ordering, and the like. What we see as brain specialization could result from the order in which we experience things in our personal development.

We know now that brain specialization and anatomy can be influenced by culture. This makes it much harder to tease apart evidence for language-specific biology from biological properties resulting from learning or the environment. For example, Carnegie Mellon University (CMU) professors Timothy Keller and Marcel Just published an article in the journal *Neuron* in 2009 showing that reading-challenged children who experienced as little as six months of intensive remedial reading instruction grew new white-matter connections in their brains. Just's study shows culture changing the structure and functionality of our brains.* Other CMU studies have shown that the connections between portions of the brain can weaken or strengthen over time, based on the cultural experiences of the individual. A similar kind of study was published as a letter to the journal *Nature* in October 2009 by Manuel Carreiras, Mohamed L. Seghier, Silvia Bagnero, Adeline Estévez, Alfonso Lozano, Joseph T. Devlin, and Cathy J. Price, in which they show that brain structure is affected by literacy and that literacy changes the anatomy of the brain.

Now, since culture can change the form of the brain and since there is no known single cognitive function that must appear in a single location in all brains, then the difficulty of using any argument from cerebral organization or anatomy for linguistic nativism is again apparent. It must not be forgotten either that, from one vantage point, localization in the brain is often the result of the trivial fact that everything that exists has to exist somewhere. If we know something, it will be found some place in our brains. This is no more evidence for innate knowledge than the fact that a famous composer comes from a specific town

---

* White matter is named for the white (because fatty) material (technically, myelin sheaths) that surrounds the nerve fibers connecting parts of the brain used for higher cognitive functions.

is evidence for predestination. Bach was born in Eisenach. Does this mean that Eisenach was predestined to be his birthplace? No. Everyone has to be born somewhere.

Human anatomy and physiology provide no direct support for the proposal that language is innate, aside from the ability of the human vocal tract to make a noise. But if the evidence within us is not compelling, it could be that there is evidence external to individual humans that might help settle the question. One set of data that is frequently adduced in support of nativism comes from the supposed similarities between the languages of the world. But these similarities are a much weaker reed for leaning an argument on than some have supposed. In fact, the most salient characteristic as we survey the languages of the world is not their similarity but their diversity.

# PART TWO:
Solutions

'Inarticulate cries, plenty of gestur
must have been for a long time the
addition, in every country, of some conventional articulate sounds
… particular languages were produced …'

Jean-Jacques Rousseau, *On the Origin of the Inequality of Mankind* (1754)

I n their letter to *Nature*, Michael Dunn and his colleagues wrote that findings from their research 'support the view that – at least with respect to word order – cultural evolution is the primary factor that determines linguistic structures' They based their arguments for this 'cultural evolution' on evidence that linguistic diversity rests on patterns of statistical tendencies. There are two widely accepted theories of the supposed correlations between features in the world's languages. For example, it is often stated by linguists that if a language has prepositions, then its direct objects will follow their verbs. So, for example, English has prepositions and its direct objects follow the verb:

*John ran to the store. We talked about the book. I gave a book to Bill.*

*I kicked the ball. Mary saw a car. Irving played the violin.*

In Japanese, however, the correlations are reversed:

*Sensei      ga      gakusei ni      hon      o      ageta.*
*Teacher (nominative) student (dative) to book*
*(accusative)   gave.*

*book to a student.'*

ample the verb, **ageta** 'gave,' follows its direct object, **hon**
oreover, the Japanese equivalent of English 'to,' **ni**, follows
ei 'student,' rather than precedes it as it would in English, so Japa-
se does not have prepositions but 'postpositions'. That is, Japanese
says 'book gave' and 'the student to' rather than 'gave book' and 'to the
student', as we order them in English. The word order of Japanese is the
most common among the world's nearly 7,000 languages.

For many years, linguists believed that such correlations were nearly
absolute and that explaining them was vital to the enterprise of theoretical linguistics. One account, based on the work of the late Stanford
professor Joseph Greenberg, explains these correlations by introducing
the idea that it is possible to make deep generalizations about the roles
of words like prepositions and verbs and that these deeper connections
explain the correlations. The other line of explanation is the notion of
'parameters' from Chomskyan theory. Children are born, according to
this theory, with the universal parameters of 'head first' *v.* 'head final' in
their genes and they look for data from the language of their environment to determine whether it is one or the other. If we take the verb and
the preposition or postposition to be the heads of their phrases, then
English is a head-first language and Japanese is a head-last language.

But the authors of the letter to *Nature* argue that neither of these previous explanations works when we look more carefully at a very large
set of data from the world's languages. What explains correlations is
whether the languages being compared come from the same linguistic family, the same cultural heritage, or not. This work is a striking
counterexample to the predictions of the universal grammar hypothesis, to the idea that language structures are in any sense innate.

Happily, this all fits fine into the idea that language is a tool. A given
culture has discovered a solution and a way of organizing its grammar.
And this solution is passed down to subsequent generations and de-
scendant languages of the original 'mother tongue.'

If languages varied little from one another except in their vocabularies, then the idea that languages emerge principally from a common
biological source would be appealing. On the other hand, the idea that

language is a tool shaped by cultural and communicative demands would make more sense if diversity rather than similarity were the hallmark of human language. When we look across the vast array of the thousands of languages spoken throughout the globe, the diversity of human languages is astounding.

At first blush, it does not seem incompatible with nativism that there should be a great deal of linguistic diversity in the world. But on closer examination, languages vary more than some 'biolinguistic' theories would lead us to believe. This point has been made forcefully in the past few years by researchers from around the world. In a 2009 article entitled 'The Myth of Language Universals,' Nicholas Evans and Stephen Levinson, researchers from the Australian National University, and the Max Planck Institute for Psycholinguistics in Nijmegen respectively, showed 'just how few and unprofound the universal characteristics of language are.' They also showed that forms in languages vary radically and thus serve to remind us that humans are the only species with a communication system whose main characteristic is variation, not homogeneity. Humans do not merely produce fixed calls like vervet monkeys, they fit their messages to specific contexts and intentions. The paper's authors also surveyed languages from around the world to make the case that the universal grammar proposal at best picks up on tendencies rather than universals.

This takes us back to an old and infamous quotation from linguist Martin Joos, often trotted out to show how much progress linguistics has made as a science since its backward past. In 1957 Joos claimed that, 'Languages can differ from each other without limit and in unpredictable ways.' Many modern linguists think that this sentiment is naive and unscientific. However, they are only partially right.

They are right in the sense that this quotation, if taken literally, would rule out the very possibility of a theory of language. If there are neither patterns nor regularities shared by the languages of the world, then there is no explanation to be found, no theory to be proposed. And it certainly does not seem that, although languages vary enormously, they vary 'without limit.' Part of the appeal of Chomsky's work is that it seems to tell us not only why Joos was wrong but also how languages differ from each other in predictable and systematic ways. Chomsky made a type of linguistic theory possible. If this new work against universals were intended to support Joos's remark, then to many scientists the paper would represent a huge step backwards. But that is not what it intends. In fact, it was not even Joos's intention. What both are trying to say – Joos aphoristically, the new research with solid evidence published in a major scientific journal – is that the differences between the languages of the world are greater than any nativist theory could predict.

This contradicts the view of what the 'Martian linguist' would find. As a reminder, according to Chomsky, '[T]he Martian might conclude that there's fundamentally one language with minor deviations.'

Nativists appear to ignore their own Martian mantra and act as though research showing the near absence of language universals were irrelevant.* They argue that even if it could be proven that there are no language universals, this would not matter for the theory of language, because the innate mental grammar they suppose is not a rigid floor-plan that all languages must slavishly follow but, rather, a toolbox from which all languages choose whatever they need or want for a variety of reasons – cultural, aesthetic, traditional, historical, and so on. This is a perfectly reasonable response, conceptually, even though it flies in the face of the Martian linguist example. At the same time, however, unless proponents of the toolbox approach can say what the tools in this toolbox actually are and why these are language-specific tools, rather than, say, general cognitive tools that are useful in multiple thinking tasks, this metaphor neither explains nor predicts anything.

When you think about it carefully, even if some nativists were correct that language diversity does not matter for their position at one level, it still creates a problem at another. This follows because even if differences between languages are all superficial, then although human languages may share many abstract properties, their actual grammars

---

* Some time in the early 1990s, I was visiting MIT and attended a large lecture by Chomsky. A group of his students were sitting in the back giggling. When Chomsky mentioned the Martian linguist example, they could barely constrain their chuckles and I saw money changing hands. I asked them afterwards what this was all about and one of them said that they had bets riding on exactly when in his lecture Chomsky would mention the Martian linguist.

are determined by a variety of other factors, a variety that turns out to be so large that we can understand the force of Joos's statement without taking it literally – there simply is no way to predict anything significant about human linguistic form if we can wave off all diversity as trivial. There is a serious problem with the idea of a priori linguistic knowledge in the light of the linguistic diversity we find in the world. The abstract similarities that the nativist is forced to appeal to, or the toolbox that some imagine, could as easily result from the fact that all languages must solve the same problems, whether communication or expression of thoughts. Under this view, similarities among languages, such as they are, would be like resemblances between bows and arrows across the world. The forms of language and bows are partially determined by their functions, via properties of the brain that have nothing to do with language or bows *per se*. The similarities could also arise from other, perhaps innate factors, such as an 'interactional instinct' – a need to interact socially and communicatively that seems to set humans apart (which is more relevant to social behavior than to linguistic behavior). And their differences might derive from the interaction of human intelligence and human environment. Alternatively, similarities among the languages of the world, such as they are, could result from monogenesis: that is, all living languages are similar because they share a common ancestor. In fact, in a 2011 article in the prestigious journal *Science*, Dr Quentin Atkinson from the University of Auckland argues that he has found evidence, based on the distributions of sounds in the world's languages, that human speech originated in a single source in Africa. If either monogenesis or the idea that the form of languages is similar because they solve a similar problem (or 'form follows function,' an idea widely accepted in biology, architecture, and other fields of learning), then the role of a priori knowledge is anemic at best.

But there *is* a problem for the nativist position from the lack of language universals that needs to be repeated and underscored: as Chomsky speaks through his Martian linguist, all grammars in the world are predicted to be roughly the same in their essential characteristics. Universals and 'sameness' have in this way always been used by nativists as evidence for their position. Evans and Levinson's article knocks these supports out from under nativism.

To drive this home, there are several issues here that are relevant for the nativism *v.* culture debate. The first of these is the matter of whether there is anything inborn in humans vital to the learning of language. Of course! Our brains are essential to learning language and those cerebral spheres grow out of our genome. Our vocal apparatus is another inborn language-relevant manifestation of our biology.

But finding genetic support for language is not the issue, though debates about nativism often make it appear as though it were. Rather, the question is whether or not there is anything *exclusively* dedicated to language learning or language form in our genotype. The answer is that we currently know of no such thing, aside from the shape and development of the human vocal tract. The final issue to think about in this regard is this: if there were something inborn and specific to language, it makes less sense to call this 'knowledge,' as some interpretations of universal grammar have it, as opposed to an anatomical or physiological feature. These are simple points yet they are at once vital and ignored in most of the vast literature on nativism. Since few, if any, of the structures that are universal to human languages or features of human biology are dedicated to language, the view that language is inborn has a far more difficult row to hoe than many realize.

The nativist refusal to be swayed by the absence of linguistic universals of any substance reminds me of the story about an encounter between a biblical literalist and a liberal theologian. The liberal passed the literalist one day on the street. As he was walking by, he jumped, startled, as the literalist shouted, 'Hallelujah! 'What are you screaming hallelujah for?' asked the irritated theologian. 'I am shouting because I just read in the Bible that God saved Moses and the people of Israel by drowning Pharaoh's armies in the Red Sea.' 'Silly man,' came the condescending theologian's reply. 'We now know that where the biblical account says "Red Sea," the translator should have said "Sea of Reeds," a shallow little swamp, maybe a couple of feet deep, nothing more.' The literalist looked at the theologian crestfallen. Smugly the theologian continued on. But then suddenly he heard the literalist once more shouting, 'Hallelujah! Hallelujah, God Almighty!' Rushing back to discover the cause of this additional idiocy, the theologian demanded, 'Now what on earth are you shouting hallelujah for? Didn't I tell you that Pharaoh's army never entered the Red Sea?' 'I am shouting,' said the literalist, 'because God drowned all of Pharaoh's army in two feet of water.' Sometimes it is hard to get people to see when they have a problem.

Once again, the problem for nativists is subtle, but critical. Remember that what is at issue is whether humans have a priori knowledge of language. If we say that people can pick and choose their language structures from a toolbox in the brain, we still have not responded to the question of whether there is anything specific to language in the brain. One could just as easily say that there is no innate linguistic capacity but that the human mind instead simply is a good problem-solver and uses some tools for a variety of purposes. We could hypothesize that in different societies the language problem is solved by similar, but non-identical methods that are not unique to language but are implicated in a range of different skills. And once we admit that, the toolbox metaphor does not leave us with very robust support for any type of a priori knowledge. This is because, even if we could show that people adopt similar linguistic structures, these could be similar solutions to the same problem, communication – like the bow and arrow solves the hunger problem and so was independently invented in many cultures – without requiring a priori knowledge of bows or arrows. And if that were the case, then the simplest solution is that the brain is a good problem-solver, not that we are born with a priori knowledge.

Nevertheless, many researchers insist that there are language-specific cognitive tools. Thus, in their famous paper 'The Faculty of Language: What Is It, Who Has It and How Did It Evolve?', which appeared in *Science* in 2002, Marc Hauser, Hauser's former postdoctoral research associate William Tecumseh Fitch, and Chomsky offered a novel proposal for what makes human language possible. They proposed to divide human abilities underlying language into two general faculties, the 'broad faculty of language' and the 'narrow faculty of language.'

According to Hauser, Fitch, and Chomsky, the broad faculty of language is the set of biological endowments that humans draw on to produce language but that are also found in other species. These, presumably, include things like the FOXP2 gene, the tongue, teeth, lungs, hands, the ability to engage in sequential planning, and so on. No one component of the broad faculty of language is unique to language, but all items of the broad faculty of language are implicated in human linguistic ability and many are shared by other species.

They propose that the narrow faculty of language, on the other hand, is not only unique for language, but that it consists of nothing but the property of recursion, defined roughly as the ability to create complex structures out of simple structures.* The contrast between narrow faculty of language, on the one hand, and Chomsky's concept of universal grammar, on the other, is striking. According to the narrow faculty of language proposal, there is only one thing in human brains that makes them differ from non-human brains with respect to our unique linguistic ability, namely, recursion. But this contradicts Chomsky's own earlier work, at least superficially, because the universal grammar proposal claims that the innate linguistic knowledge of humans is far more extensive and detailed than merely recursion. Highly specific principles, rules, and constraints – lists of items numbering in the thousands potentially – are proposed in the literature of Chomskyan theory.

If we are to get a handle on how evolution has prepared humans for language, we need to think more carefully about the narrow faculty of language and the broad faculty of language, on the one hand, and universal grammar, on the other, assuming that the two sets of claims go together. But there are several problems with both proposals that should engender a healthy skepticism from the outset. The simplest way to state the problem of universal grammar is that the theory is more elaborate than is needed to account for language. It is ironic that in the nativist literature there is a strong desire to achieve a biolinguistic perspective, yet there are few, if any, serious discussions of biology in this work.

Before going any further, though, we should consider what this a priori knowledge of grammar is like in more detail. If *Homo sapiens*

---

* One definition of recursion is a process that applies to its own output. A different type of definition is that a recursive structure is one in which one category contains a category of the same type. An example of the latter is 'Bill said that John would be OK.' In this structure, the larger sentence, 'Bill said that John would be OK,' contains a smaller sentence, 'John would be OK.' There is thus one sentence inside another and so the entire sentence is a recursive structure.

were born with a 'language organ,' then it might not be considered unreasonable to ask where this organ resided. But, as I have just discussed, there is no area of the brain that could serve as such an organ. Yet, linguists like Stephen R. Anderson and David W. Lightfoot, authors of the 2002 book *The Language Organ*, claim that there is literally an organ dedicated to language in the human body. This is not intended as a metaphor. I discussed this work in a 2005 journal article, from which some of the following material is taken.

It is moderately curious that this language organ cannot actually be seen, however. No point in even looking for it. Those who believe in a language organ seem to have one or both of two meanings in mind, a mental organ and a physiological organ. The mental view of the language organ takes it to be formed by knowledge of two types, the a priori type and learned knowledge, gained at the final stage of linguistic development when a child achieves mastery of the adult grammar. This is a popular idea that has gained widespread purchase. But despite the popularity of this conception of the language organ, there is a problem with referring to knowledge as an organ. It has unwanted implications, some leading to absurdities. Philosophers contend that knowledge is 'warranted belief,' that is, belief that has a basis in more than just blind faith. But if this were so, then by the same reasoning, if language knowledge were an organ, all of our other knowledge and beliefs would also be organs. This does not make sense to me. And no doubt proponents of a language organ would agree. If we conceive of knowledge as an organ, we reach an absurd conclusion.

But a second use of the term 'language organ' is that it is physical in some way. In their book, Anderson and Lightfoot suggest that this organ could be tissue that is specialized as a result of an animal's biological organization, tissue that is sensitive to certain environmental events. In this view the language organ is 'a biologically determined aspect of certain tissue (primarily in the brain), rendering it uniquely sensitive to linguistic events in the environment.'

But even this reasonable hypothesis faces problems. First, if the language organ is nothing more than 'specialized tissue,' is it not fair to ask where in the brain this tissue is found? The answer is nowhere and everywhere. Tissue underlying language tasks is found all over the brain and even outside of the brain. Is the tongue part of an overall language organ, for example? Or is it one of many language organs? Of course, even the tongue has multiple functions, language being but one. You can stick your tongue out to express defiance or disrespect; it moves the food around for mastication; it cleans your teeth; it tastes your food; it makes your lover happy. Is it also a food organ? A sex organ? An organ for any use to which it is put? Not only is it not clear whether there is a language organ as 'specialized tissue', there is a deeper problem with talk about specialization for language in the brain. Neuroscientist Professor Friedemann Pulvermüller of the University of Cambridge, for example, makes a convincing case that neurolinguistic studies of syntax are unable to prove that the activities they observe in the brain are strictly related to the linguistic behavior correlated with this brain activity. As he puts it in his 2002 article for *Progress in Neurobiology*, 'A Brain Perspective on Language Mechanisms, '[T]here would not be a high correlation between a bold brain response and a well-defined linguistic process, but there would be a brain response co-occurring with a variety of linguistic and psychological phenomena.'

What Pulvermüller is saying is that claims about language organs are premature, perhaps misguided altogether. But even if we ignore the caveats of leading neuroscientists, problems remain for the idea of a language organ. Let's say that talk of specialized tissue might be warranted by neuroscience. There is still a problem because the definition given specifically avoids saying that the language organ is found in a specific spot. Rather, it is a system of dedicated tissues, neurons, etc.

But what follows if this organ has no specific location in the brain and is merely the dedication of a certain portion of brain tissue (from an area that might allow other 'dedications' as well, since we know that various parts of the brain are multifunctional)? By this definition is Broca's area an organ? Or is my knowledge of recipes a 'gourmet organ', if that knowledge can be located in a specific part of my brain? *All* human knowledge is found in the human brain and in some sense the tissue in which the knowledge is contained is specialized for this knowledge. But this tells us little about the ultimate source of that knowledge.

Considerations like these make it difficult to talk about a language organ in our present stage of neurological knowledge. However, maybe there is a way to forgo talk of a language organ, while holding on to the related ideas of language instinct and universal grammar. In fact, it would be possible to separate these other concepts logically and conceptually from the 'language organ' and to reject the latter but keep the others. The problem is that all of these ideas face basic problems equally as daunting as the language organ proposal.

Scientific hypotheses have to be evaluated rigorously. One of the most important ways to evaluate a hypothesis is to test its predictions. For example, the universal grammar hypothesis predicts that normal, healthy individuals from all manner of human backgrounds and experiences will always reach the same 'final state' of knowledge of the grammar of their language. This is an interesting prediction. After all, if everyone has different experiences, how can they all end up at the same knowledge of their grammar? This surprising state of affairs would not only be explained by universal grammar, but universal grammar actually says it must be the case. The problem with this prediction is that it has never been tested. We just do not know if all language learners actually do wind up at the same 'final state' or not.

Universal grammar researchers have, to their credit, however, proposed an interesting method for investigating whether all speakers achieve the same knowledge of their grammars. The idea is that we should take sentences that are rare, that most native speakers will have had little or no experience with, and, varying the sentences systematically and minimally, test to see if native speakers have the same judgments about whether such utterances are grammatical or not.

I have in fact attempted to do this many times with Amazonian tribal groups. None that I know of have words for 'grammar' or 'grammatical' so I test their judgments by asking them whether or not this or that sentence is 'pretty' or if they would say it or not. Sometimes the results are unexpected. One of my most helpful language teachers insisted that I could say a certain sentence and that when *I* said it, it was indeed 'pretty.' However, I asked him to say it himself, having learned that unless the native speaker will utter the sentence there is probably something wrong with it, no matter what they say. He replied, 'I cannot.' 'Why not?' I asked. 'Pirahã do not talk like that,' was the puzzling reply. 'But you said I *could* say it! Yes,' he said, 'you can say anything you like. You are paying me.' This exchange showed me that I would have to use more sophisticated methods to test native speaker knowledge. Although it does not, of course, rule out looking for consensus of judgments in principle.

Let us assume that all the problems can be addressed and that some method of evaluating speaker knowledge can be used satisfactorily. The prediction is this: any two native speakers of a language should be able to produce identical judgments of whether potential English sentences are grammatical or ungrammatical. For example, if someone utters, 'Who do you think that John saw?', English speakers from the same community should agree that this is a grammatical sentence. If they do not, then the prediction is wrong. However, their judgments should be the opposite about the very similar example: 'Who do you think that saw John?' On the other hand, if we then remove the word 'that' from both sentences, then both should be hunky-dory for all native speakers, as in 'Who do you think John saw?' and 'Who do you think saw John?'

This is not the only kind of evidence marshaled by proponents of innate language to show that we all acquire our languages equally well. But their prediction is not obviously wrong. There are many highly intelligent and competent linguists, philosophers, psychologists, and others who accept it as well established. Yet it is curious that no nativist claim for homogeneity of language acquisition grapples with humans' relative abilities at story-telling, writing poetry, using language to woo a lover or to win over enemies. One reason that these abilities are avoided as a testing area is because they definitely do vary from one speaker to another. We may all know how to make a noun phrase, but we cannot all write good novels. If these skills are part of language, then the different abilities are not predicted by nativists.

I am not done with this subject, but I want to detour in order to make one point about the idea behind the universal grammar or language organ that some structures which are known equally well by all native speakers share a particular grammar. Whenever someone tells you that their scientific idea makes a prediction, ask yourself three questions. First, 'Does it really?' Second, 'How good is the evidence?' Third, 'Is there a simpler explanation?'

Universal grammar predicts a degree of homogeneity of judgments by native speakers as to whether this or that utterance is 'good' or 'bad'. There should be groups of speakers who share identical syntactic principles, and therefore judgments about what is grammatical. Then we must ask whether this is correct. If it does turn out to be correct, then another question arises, namely, whether people were born with this knowledge or learned it. Anyone who wants to propose that knowledge of language is inherent in the species is obliged to answer these questions.

The answers to date are not favorable to the idea that speakers' judgments are particularly homogenous. I have taught syntax regularly, to Portuguese speakers in Brazil, and to English speakers in the USA and England. As I polled native speakers of English and Portuguese among my students, I rarely found anything approaching 100 per cent agreement on the grammaticality of examples used in classes, whether of my own invention, from an article, or from a textbook. And this included people who came from the same neighborhood, social class, and gender – people you would expect to share a common dialect (and who so described themselves when asked). It might be, though, that these results of mine are of no real significance because they are insufficiently nuanced. Other researchers have made the convincing case that if linguistics were to adopt the statistically based methodology of standard social science, then it would be possible to test the degree to which speakers might be said to share grammars more effectively. For some reason, most linguists have failed to implement on a large scale more rigorous testing of the ideas of a priori knowledge, ignoring methods that have long been used by psychologists and sociologists. Whatever the reasons, the upshot is strange – the hypothesis that very specific linguistic knowledge is found in our genes remains relatively untested.

On the other hand, if we take some nativists literally, there is really nothing to test. Universal grammar, the language organ, the language instinct, and so on are all necessarily true. If only humans have language, then human biology must underwrite language. And so whatever our languages look like, since they are only possible due to something in human biology, this biology is itself the universal grammar. That is not a satisfying answer, however, because all agree that only humans can use human tools well. The discovery that humans are better at building human houses than porpoises tells us nothing about whether the architecture of human houses is innate. But that is the crucial question nativism must answer: Is there is anything language-specific (or house-specific) about human biology?

Some say that Wernicke's area is a linguistic part of the brain. To be sure, it is demonstrably important for certain linguistic tasks. But, as we have seen, the evidence suggests that the tasks Wernicke's area is responsible for are much larger-grained than the merely linguistic – whatever linguistic and other Wernicke-related activities have in common, sequentiality, for example. Looking for abilities that generalize across the various tasks that this area is supposed to be responsible for is a more perspicacious way of describing Wernicke's area than to say that it is 'for' any of the more specific abilities associated with it.

A serious weakness in the hypothesis that there is innate knowledge derives from a failure to consider alternative explanations. One thing that human beings are good at is learning from other human beings. We cannot talk about what makes us human without talking about our learning ability. And there *is* an organ for this. It is called the brain. Humans have better brains for learning than other animals. Which is innate. But that is not the same as a priori knowledge. This is not to say that the brain does not have specialized regions and abilities, such as those used for vision, hearing, smell, and so on. Nor that language cannot be at least partially underwritten by brain regions that evolved for other things. But general human intelligence is still a major reason why we can learn language. It is vital, in fact, because human language is so flexible and variable across the world or even within specific communities.

Such failures to find convincing corroboration for nativism undercut innatist ideas on grammar. They are more elaborate than we need, violating the well-known principle Ockham's Razor, named after the fourteenth-century Franciscan philosopher William of Ockham, who proposed, '*Entia non sunt multiplicanda sine necessitate*,' 'Don't multiply entities beyond necessity.' The great twentieth-century philosopher Bertrand Russell interpreted this to mean: '[I]f one can explain a phenomenon without assuming this or that hypothetical entity, there is no ground for assuming it, *i.e.* that one should always opt for an explanation in terms of the fewest possible number of causes, factors, or variables.' We cannot explain human cognition without referring to the brain. The question is whether we need to posit anything more than the brain to explain language. In particular, do we need a universal grammar, language instinct, language organ, or whatever we call it? If we can explain human language without this additional type of entity, then, by the principle of Ockham's Razor, we should. If language diversity in the world can be explained without any of these entities, then this is one less reason to propose them.

A nativist should also be troubled by an interesting prediction of nativism that has never been seriously tested: that there are genetic mutations for language differences. This prediction was first pointed out by Brown University cognitive scientist Philip Lieberman and it is worth dwelling on a bit.

We know that genes can develop or mutate very quickly. We also know that culture often plays a role in genomic changes. Consider the following extract from a story in *The New York Times* of July 1, 2010:

> *Comparing the genomes of Tibetans and Han Chinese, the major-*
> *ity ethnic group in China, the biologists found that at least 30 genes*
> *had undergone evolutionary change in the Tibetans as they adapted*
> *to life on the high plateau. Tibetans and Han Chinese split apart as*
> *recently as 3,000 years ago, say the biologists, a group at the Beijing*
> *Genomics Institute led by Xin Yi and Jian Wang.*

Now consider the fact that in many versions of innatist theory, different languages may, by hypothesis, set different 'parameters.' A parameter is a kind of 'switch' that we are, by this view, born with. One such is the 'invisible subject' parameter, illustrated in both its values in Portuguese (which allows invisible subjects) and English (which does not), respectively: '*Vem amanha*,' 'Comes tomorrow.' In Portuguese this can mean that he, she or it will be coming tomorrow. But in English it is claimed by proponents of the invisible subject parameter that the equivalent to the Portuguese sentence is not possible, that is, that we cannot say, 'Comes tomorrow' as a grammatical sentence. The idea that English prohibits invisible subjects – equivalent to saying that most sentences in English must have a subject that can be heard – becomes especially interesting when we remember that English has subjects that do not mean anything. These are the subjects 'there' and 'it.' Linguists label them 'pleonastic' subjects. In English we can say '*There* is a man in the room' or 'A man is in the room,' but we cannot say (except as a question) 'Is a man in the room.' And we can also say, '*It* is raining' (what does the *it* there refer to? *What* is raining, for heaven's sake?) but never, 'Is raining.' If English may not have invisible subjects, then every sentence will need a subject we can hear (or see on the page).*

Right, so that is the invisible subject parameter. Let's grant for a moment that the innatists might be correct that this parameter is genetically specified. The question then is why there are no societies known that have evolved a genetic *in*ability to learn a language with invisible subjects. One could imagine that some society not only lacks invisible subjects but that its members are genetically incapable of learning such a language. Take a child with such a genetic difference and raise it in Brazil and they could never learn Portuguese. Or imagine the opposite – some group evolves a new gene that prohibits all but invisible subject words. They would be unable to produce or to understand subjects that are pronounced, like those of English. These mutations are representative of the *overlooked* theoretical predictions made by the theory that grammar is innate. On the other hand, these types of genetic mutations are *not* predicted if grammar is learned by general intelligence in humans because then there is no genetically based parameter of null subjects, but only that universal possession of all *Homo sapiens* – intelligence.

Continuing along this line of reasoning, it is interesting that no one

---

* This invisible subject parameter is poorly worked out, however, at least when we consider English discourse segments, which indicate that English also allows invisible subjects. My friend David Brumble, Professor of English at the University of Pittsburgh, underscored this point in a letter to me in which he remarked, 'Consider a desolate stage. A man stumbles into it and looks around with alarmed eyes. He mutters to himself: "Am scared. [Pause.] Am supposed to be scared, eh? [Pause.] Am about to be attacked maybe? [Lies down.] Will pretend to be dead. Safe now."' I think that this sequence is possible. If so, then English – and maybe all languages – allow invisible subjects under the right circumstances. The failure to consider such examples is a result of failing to consider discourse as part of grammar – a typical perspective of nativists, for some reason.

has ever shown that languages are incapable of being learned. Why is this important? Well, if grammar is innate then it is not learned. In fact, some have claimed that grammars are grown like any other organ of the body. Not only are they not learned, according to this reasoning, but *they cannot be learned*. There are entire shelves of books and articles with titles like 'The Logical Problem of Language Acquisition' (2009) edited by C. L. Baker and John McCarthy that work to demonstrate this. Each of these works claims that grammars cannot be learned, therefore they must be innate. But for each book that concludes that language cannot be learned, it seems that another book exists that decides the opposite. In recent years, for example, Steven Pinker's 1994 book *The Language Instinct* was in principle answered by another, *Constructing a Language* (2003), in which the author, Michael Tomasello, argued not only that languages are learnable, but that there is nothing particularly remarkable about how they are learned, compared to other things animals learn.

In fact, the evidence is overwhelming that language is learned, rather than 'grown' homogeneously by all normal members of the species. Language in every society requires years of experience and exposure to data for any child to reach adult levels of fluency. These are hallmarks of learning, not of genetic determinism.

Likewise, there is no compelling evidence that all people learn language equally well. If language is innate in the same sense as growing an arm or hair is innately specified, then it makes no sense to claim that some people are better at learning language than others, just as it makes no sense to say that some are better at growing hair, aside from cultural preferences, such as 'Johnny's hair is thicker and prettier than Joanna's' or some such. This does not mean that Johnny is any better at growing hair than Joanna, only that members of his culture like the way his hair grows, independently of any ability of his. And the same is true for language. If language were innate, then intelligence, social class, economic opportunity, and so on would be largely irrelevant for whether or not an individual mastered their native language as well as any other individual. This is a crucial prediction of nativism. For if people could be shown to have learned languages at significantly different levels of ability, then not only would it make more sense to say that they are using general human intelligence to learn language but the mastery of their native language would be more like bicycle riding than growing body parts.

So far as I know, however, this prediction has never been convincingly tested. Rather, it is taken as an obvious fact. Maybe people do grow languages. Maybe there is no significant difference in the level of mastery attained. But the prediction is so vital to the nativist model that it is surprising to learn that there is no experimental support for it. In fact, the hypothesis is difficult to test in principle. How on earth *could* we tell whether Billy and Marty speak their native language equally well? What criteria could we choose that would be free from observer bias? What structures would determine this?

Imagine that you are in the jungle living among a tribal group, such as the Banawá in Amazonas. A few men are sitting around a fire, under a thatched roof in a small hut surrounded by jungle. All are sitting just outside the full glow of the flames, talking about their day's hunt. Each tells a story in a slightly different way from his fellow Banawá. One uses one kind of syntactic construction more than the others. Another uses more nouns. Another uses certain verbal suffixes more than his friends. How could we tell whether one of them speaks the language 'better' than the others? We could ask them. But that will get you as many different answers as there are men in the room. We could try to analyze their speech to determine who speaks the best, but it is not likely that we would discover any good criteria to do so. It is possible to tell whether one person speaks their language better than others. But this is a very hard task which requires more types of information than the average linguist is likely to have.

Likewise, it has never been established that people do not have sufficient data to learn language. This is vital to show if you believe that language is innate. Everyone agrees that if something is learned it is not innate. Plato and Socrates believed that we know things we do not learn and cannot have learned in this life. Nativists believe this too and they call it the 'poverty of the stimulus' argument. The idea is just this: figure out what people know about their language, grammar, etc., then figure out what they could not have learned. Subtract the difference. The result is what is innate. That sounds like a simple process. But it is worthless unless one is able to show just what it is that people could not have learned. And yet several writers have shown that the idea that the stimulus is too poor to account for learning has serious problems.

Let me be clear that no one has proven that the poverty of stimulus argument, or Plato's Problem, is wrong. But nor has anyone shown that it is correct either. The task is daunting if anyone ever takes it up. One would have to show that language cannot be learned from available data. No one has done this. But until someone does, talk of a universal grammar or language instinct is no more than speculation. The stimuli that all healthy humans are exposed to are many and rich, likely sufficient to account for how people can learn, rather than 'recall,' language.

Nativism also assumes that there are stages of human development specific to language. But no one has ever demonstrated this either. Although many researchers have done a wonderful job of laying out what we know about how language develops in the child, no one has demonstrated that the stages of development that children pass through in their linguistic experience are not just part of more general developmental stages that have effects in many other domains of cognition. Let us say that all children in all cultures around the world pass through the same set of developmental stages as they acquire their native language and that they pass through these stages at roughly the same time in their lives. Does this show that language is innate? Not at all. Its brain needs to mature in many ways before any human infant can perform well at adult human cognitive skills. These different stages in principle have nothing to do with language *per se*, but rather reflect developmental plateaus in the larger-picture development of the whole brain. This is the proposal that the large number of computer scientists and psychologists working within the broad framework of connectionism offer, based on their research.

Scientists at Carnegie Mellon University, the University of California at San Diego and other institutions in various parts of the world have shown, using a number of simulation studies, how computers that learn merely by statistical models duplicate the developmental stages of human children. But of course computers do not have a language instinct or a language organ or anything of the sort. Thus the researchers show that if children have the innate general ability to make generalizations based on statistical associations between elements in their environment (that is, if the children are roughly like little Google search engines instead of possessing little cookie cutter language molds in their brain), then all of the stages of development that have been used to illustrate the power of the language instinct show only the larger power of the human brain as a general-purpose computer and the stages of learning it, logically, must pass through.

Based on the evidence we have been discussing and our present state of knowledge, the hypothesis that language is innate makes no uncontroversially established predictions that would support it over the alternative that important principles are learned by experience via an all-purpose cerebral computer. We know that we have to allow some general-purpose computing power to the brain. But then Ockham's Razor rises into view because the nativist attempt to solve Plato's Problem runs the risk of multiplying entities beyond necessity. It offers an explanation richer than what we actually need.

If general intelligence is all there is to it, then why do all humans speak languages and why do those languages share any similarities at all? If there is no language instinct or universal grammar, then why do all people learn languages? Why don't some people, perhaps some of lesser intelligence, remain mute, failing to learn grammar at all, no matter how hard they try? Well, in the first place, general intelligence is *not* all there is to it. There are areas of the brain that, because of their general properties, also constrain and enable language, though none of these is specific to language – they house abilities that not only language but a range of other cognitive abilities must draw on, such as sequencing the parts of complex actions like riding a bicycle or combining words into phrases. There are several possible explanations for why we all speak and why languages are similar in many ways. Some of these possibilities have never been seriously considered. Perhaps the easiest and best hypothesis is monogenesis – language evolved only once in human history, in Africa, before the great *Homo sapiens* diaspora. Another is that humans share general learning abilities and all humans have to solve the same communication problem – this would predict that humans would independently discover many of the best solutions to this problem, leading to similarities between human languages. If that were true, then it could be that all the similarities between human languages follow from the common ancestor or common solutions to common problems, and not from an innate universal grammar.

Whatever our thoughts on the likely origin or relationship of language to culture or human cognition generally, we need to think in more detail about the mechanics and purpose of language. One of the best ways to approach this is to ask the question: How would you go about building a language?

## Chapter Six

# HOW TO BUILD A LANGUAGE

*'It is the pervading law of all things organic and inorganic, of all things physical and metaphysical, of all things human and all things super-human, of all true manifestations of the head, of the heart, of the soul, that the life is recognizable in its expression, that form follows function. This is the law.'*

Louis Sullivan, 'The Tall Office Building Artistically Considered' (1896)

Louis Sullivan was a famous American architect working in the late nineteenth and early twentieth centuries, known in part for his influential and controversial idea that 'form follows function.' The controversy over his ideas arose because many subsequent architects, artists, and other thinkers interpreted Sullivan to mean that buildings, tools, and other artifacts should have no form other than that which is necessary to fulfill their functions. This led to a puritanical abhorrence of ornamentation for its own sake. It was as though everyone believed that clothes should be neither stylish nor colorful, but only warm and waterproof.

This view of ornamentation is incorrect. And I do not believe that it necessarily follows from Sullivan's words. Aesthetics are not irrelevant. Ugly things, like flat roofs – which often leak – can be dysfunctional. But beauty in itself can be functional, as Darwin showed via his evidence for sexual selection – roughly, one sex trying to attract the opposite sex for the purpose of mating – as an evolutionary pressure parallel to natural selection. People don't mate with just anyone whose plumbing works. They look for beauty, among other things.

In other words, humans and other animals are obviously attracted to one another because of ornamentation and non-functional facets of the other's phenotypes. And yet because beauty has the result of giving people with certain body mass indexes, symmetrical faces, large shoulders, or bulbous mammary glands a greater chance of producing viable offspring than those of us who lack such ornamentation, it is clear that beauty and ornaments can be function-driven features of form.

There is of course no literal *sexual* selection of architectural designs. But if you want your building fully occupied, then, in addition to structural form following building function, you will need to include forms shaped by cultural knowledge to be beautiful, in order to attract the eye and pride of potential and current tenants.

In language we find a similar situation. Most forms in languages fit the function of communication without concern for aesthetics. Stories and conversations in day-to-day life simply work to get a message to a reader or a hearer. Yet poetry, novels, onomatopoeic words, and the like incorporate beauty and attractiveness as part of their functions.

When utility is the only driving pressure for linguistic forms, we could say that the resultant items lack ornamentation. Speaking loosely, words are forms for concepts (to run) and objects (dogs), phrases are forms for modified concepts (runs quickly) or objects (big dogs), sentences are forms for thoughts, and groups of sentences, paragraphs, or discourses are forms for larger expressions of multiple thoughts, usually centered on a particular theme.

But one might ask in what sense the shape of words is the result of their function. Linguists and philosophers have long known that the relationship between the form of a word and its meaning is largely arbitrary. There is nothing inherent in the sounds or length of the word 'earth,' for example, that connects it to the meaning of 'the planet humans live on.' That link is one of convention only.

Some aspects of the shapes of words and other linguistic items, however, *are* partially determined by their functions. We can see this in sentences like:

*John gave the book to Bill.*
*John wrote a book about Bill.*

The preposition 'to' is shorter than the preposition 'about.' This could be a mere coincidence or it could be something deeper. The late Professor George Zipf of Howard University formulated an explanation of the relative length of words that has come to be known as 'Zipf's Law'. His law predicts that more frequent words will be shorter than less frequent words. Therefore, because the preposition 'about' is less frequent than the preposition 'to', 'to' is shorter. Being shorter makes this preposition easier to produce than 'about', lessening the total effort required in language production, because of its more frequent appearance in speech. Zipf called his law the 'principle of least effort'. If correct, Zipf's Law would clearly support the idea that form follows function. By and large, it does seem to hold, though recent work has given a new twist to it.

Researchers from the Department of Brain and Cognitive Sciences at Massachusetts Institute of Technology have concluded that the functional motivation for word length in any language is based, not on the amount of physical effort the word requires, but rather on the amount of information the word carries. This could alternatively be thought of as the amount of mental effort the word requires, rather than the articulatory effort, as Zipf seems to have been thinking.

The preposition 'to' does not carry much information. It does little more than distinguish the indirect object of a verb from the direct object. In many languages, including English, it is not even required: 'I gave the book to John' v. 'I gave John the book.' In the Wari' language of Brazil, there is in fact no preposition equivalent to 'to.' The Wari' would not be able to say 'I gave the book to John' but only 'I gave John the book.'

The preposition 'about', on the other hand, tells us more concerning its object. In a sentence like 'I read the book about John,' 'about' tells us not only that its object, 'John,' is not the indirect object of the verb, but also that 'John' is the theme of the book. Since 'about' gives more information than 'to,' then 'about' cannot be omitted. This explains why we cannot say 'I read John the book' and mean by that the book was about John. Too much information is lost if we leave out 'about,' as we did with 'to.'

The correlation between length and information does not stop with words, though. The longer the sentence, the more information it communicates. Likewise, a longer book contains more information than a shorter book. That is not to say that more information means better information! A long book may contain a good deal of useless information.)

Knowing that the length of a word is determined by the amount of information that the word contains is a striking source of support for the idea that linguistic function shapes linguistic form. At the same time, words and sentences have more functions than merely to express definitions.

If I overhear someone say, 'I just love that lavender skirt on you,' I will know, as a member of the culture in which this sentence was uttered, that both the speaker and the recipient were likely female. None of this gender information is part of the definitions of the words themselves. Added to the literal meanings of the words in every utterance is further information about culture and society. At one level *all* the meaning of every word is cultural, because the existence of a word in any language, from prepositions to verbs and nouns, is a cultural decision. Psycholinguistic studies show that even learning the meaning of prepositions like 'on,' 'in,' 'at,' 'about,' 'in front of,' and so on requires intense cultural exposure and learning.

Three of the many Amazonian languages I have conducted field research on – Pirahã, Banawá, and Wari' – each have only a single preposition. In languages like these, prepositions are not used to distinguish concepts like 'with,' 'on,' 'about,' and so on. The languages instead use verbal suffixes or other devices, in conjunction with their single preposition, to express these meanings. Their single preposition tells the listener to pay attention to the verb to see what exactly is meant.

And so, cultures can put specific kinds of information in various places. The important thing is that if the information is important to the speakers of the language they will come up with a way of expressing it.

Researchers at the Max Planck Institute for Psycholinguistics in Nijmegen are world leaders in the study of the effects of culture on language and vice versa. They cite work on the Nisgha language of Canada, which has simple words with meanings like 'on something horizontal,' 'concave side down,' 'flat against,' and so on. Where someone speaking English would say, 'The cat is on the mat,' the Nishga might say, 'The cat is mat *flatagainst,*' and so on, with one

word for the two English words 'flat' and 'against.' That such relatively complex concepts, from an English speaker's perspective at least,
appear as simple words shows us a couple of things. First, it makes
it clear that meaning and the form that it takes is partially cultural.
We do not have individual words that mean these things in English.
Our culture has not made such meanings basic. Second, what counts
as 'more information' is also cultural. What is expressed as a phrase
in English, for example, is a simple word in Nishga. This is interesting because the decision by Nishga speakers to encode these meanings into single words means that this type of information is more
frequently used, more culturally common, and thus less informative
than the same information would be in English.

Cultures add their individual functional dimensions to form. In
addition to words, other examples come from jokes, poems, stories,
idiomatic expressions, and so on. We encounter two form-building
parameters at work in words and sentences. The first is the general
principle deriving from the very nature of communication, namely,
that more information (in the context of a specific culture) = longer
linguistic unit. But then we also have the principle that 'culture determines information complexity' or, to put it another way, 'what counts as
simple in one language may count as complex in another.'

If we were going to set about building a grammar, what would we
need to do? It does not matter whether we are using this information to
arrive at a theory of some hypothesized portion of the human genome
dedicated to grammar or merely telling someone else what they would
need to do in order to construct a human language. Whatever our theoretical position, this is a hard question. It is so hard that, in spite of the
centuries that scholars have spent trying to understand language and
grammar, no one has answered it. But they have come up with some
good ideas. There are the building blocks of language that we can all
recognize. I got my first significant look at some of these components
of human language before I entered high school. English grammar was
a highlight of those early school days. There was just something interesting and fun about the Reed-Kellogg sentence diagrams that Ms
Bartholomew taught us in eighth grade English.

It all made sense. Nouns, verbs, prepositions, adjectives, adverbs

## Figure 3 **Reed-Kellogg Diagram**

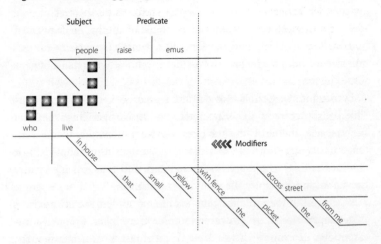

and the phrases that they cloaked themselves with were things of beauty. Many a pleasant class hour was spent engaged in figuring out how to diagram sentences like this one: 'The people who live in that small yellow house with the picket fence across the street from me raise emus.'*

I know now that the diagrams that were taught at that time are simply teaching tools and not a theory of English grammar. But I like them none the less because they challenge students to think of the structure of their utterances, helping them to reflect on the composition of phrases they read working the muscles of their minds. There are other linguists today who owe their scientific interest in language to the efforts and instruction of fine elementary school teachers.

Still it is worth asking whether these diagrams are real in either a psychological or a cultural sense or whether they are merely a game for eighth graders. Can all languages in the world be diagrammed in this way or only English? What generalizations can we make about linguistic structures that will be valid across all or most languages of the

---

* This diagram can be found at http://www.utexas.edu/courses/langling/e360k/handouts/diagrams/diagram_basics/basics.html

world? We can only answer these questions if we figure out what a language is, what its components are, and what functions it serves.

To do that, however, we need to understand the mechanics of human language. We need to determine what components are required to build a language. And we need to know what this language would be for.

Looking at the Reed-Kellogg diagram again, it is striking how sensible the structure of sentences appears – structure follows meaning fairly closely. Therefore, the first questions we must answer for any sentence whose structure interests us are: What does it mean? What is it communicating? The diagrammed sentence above provides information about some people. The main idea transmitted is that these people raise emus. But the sentence contains more than just the information that 'people raise emus.' However, in order to decipher it properly, it is essential to determine what each word and phrase is contributing to the meaning and purpose of the sentence as a whole. When all is said and done, all of us already know these complicated things at some level, or we could not understand the sentence in the first place. And yet it is often very hard to talk about that common knowledge.

In this example, the first word, 'the,' tells us that the identity of the following noun, 'people,' is known to both the speaker and the hearer. 'The' marks 'people' as definite – which means that this is a specific set of people, either personally known to the hearer or previously identified in the conversation. Following the noun 'people' an entire sentence intervenes before we come to the verb 'raise.' This sentence is 'who live in that small yellow house with the picket fence across the street from me.' This is a 'relative clause' and its function is to remind the hearer that they already know who these people are, and that they share this knowledge with the speaker. The relative clause just says in effect, 'Hey, remember those guys?' It clears up any confusion as to which people are being talked about.

We thus have a sentence, the subordinate relative sentence, inside the main sentence, 'The people raise emus.'* Following this sentence is

---

* This is an example of recursion, by the way. It is recursive because one sentence *recurs* inside another.

the main verb, 'raise,' and the object of 'raise,' 'emus.' There are several divisions in this main sentence, all represented perspicuously in the diagram by full lines and half-lines.

The first division that we see in the sentence diagram was first recognized by Aristotle, over two thousand years ago. It is between the *topic*, or subject – old information already known to the hearer – and the *predicate*, which provides new information about that topic. Most main sentences in English are built like this. In fact, most sentences in most languages begin with the topic, following this with new information that the hearer presumably did not previously know about this topic. For example, in the diagrammed sentence, the hearer is being given the new information that 'emu raising is done by a group you know of, the people with the yellow house and picket fence across the street from me.' The people with the yellow house are known to the hearer. That they raise emus was not, however. That is why the speaker uttered this new sentence. The new information-bearing predicate, 'raise emus,' is itself divided, shown by the half-line in the diagram, between the verb, 'raise,' and the direct object, 'emus.' The object noun, or phrase when there is more than one word, can follow the verb, as it does in English, or precede it as it does in Japanese.

So far each of the parts of the main clause is clear and their ordering is the result of a set of simple choices made by individual languages, such as 'place to the left of the noun,' 'place to the right of the verb,' or 'place topic to the left of predicate,' and so on. This is not very complicated.

Within the subordinate relative sentence, the principles of ordering and function are the same, except that the relative pronoun appears at the front of the sentence. In most languages the relative sentence follows the noun it is referring to, but it could precede it. The possibilities are numerous. But they are far from infinite. Scientists are not mainly interested in counting things in any case. We want to understand the principles that tell us how sentences are put together. The principles that make the most sense are those based on the meaning, efficiency, and style that word orders and sentence structures support and signify.

Culture and information are not the only functional pressures that mold the forms of language, however. There have been several interesting studies which show that the ability to analyze sentences upon hearing them – called 'parsing' – plays a large role in determining the order in which words and phrases occur in sentences in different languages of the world. These studies account for sentence form partially in terms of cognitive utility and efficiency. Reed-Kellogg diagrams represent graphically some of the meaning relationships and ordering decisions that form a sentence. And we are beginning to see that most ordering decisions are made to fulfill functions. They are utilitarian. For example, another general rule of thumb in natural languages is that words that are closer in meaning will usually be closer to each other in the sentence, though there are exceptions to this. We say, 'The black sheep ran to town,' not 'The sheep ran black to town the' because, in most languages, we expect adjectives and articles to be close to the noun they modify.

There is no single principle that accounts for all facts about sentence structure. But the list of principles that are needed to explain this structure is not large. The task for the child learning their native language is not nearly so difficult as some seem to think.

One of the many research reports on the principles of word order comes from the laboratory of Professor Edward Gibson of MIT's Brain and Cognitive Sciences Department. Gibson's lab has looked at a well-known fact about languages around the world, namely, that most languages use the basic order of subject, then object, then verb to organize their sentences. English and many other languages usually use a slightly different order – subject (S) first, verb (V) second, then object (O). To appreciate these important facts, compare how English and Pirahã both would say 'The man ate meat.'

> *English (SVO): The man (S) ate (V) meat (O).*
> *Pirahã (SOV): xigihí (S) sigíhií (O) koháipí (V).*
> *'man     meat        eat'*

Why is the SOV order of Pirahã more common in the world's languages not only than the SVO order of English but than any of the other possible orders, such as VSO (eat man meat), VOS (eat meat man), or OVS (meat eat man)? VSO is often found in different parts of the world, including in classical Arabic and Welsh. The word orders OVS and VOS are very rare, but I myself have written a 540-page grammar of the Wari' language of western Brazil, which is a VOS language. OVS to date has only been found in one language of the world, Hixkaryana, another Amazonian language, spoken several hundred miles north of the Brazilian city of Manaus.

Gibson's research showed that the order SOV is not only the most common order encountered in real languages around the world, but that it is also the preferred order when we gesture – even when our native, spoken language is not SOV. English speakers, for example, used SOV order for gesturing in experimental conditions, even though English is SVO. If you ask a native speaker of English to describe a scene with gestures, Gibson's team showed that they will make a gesture for the subject, then one for the object, then one for the action – SOV. If SOV is indeed the preferred order for arranging sentence parts among all humans, then why are there any other orders at all? Why do English and Spanish both use the order SVO? The research group proposed that the high percentage of SVO languages results from communication pressures over a noisy channel.

But before explaining this last line, let us consider one explanation for why the majority of human languages prefer to put the subject before the object. The subject goes first in most languages because it usually corresponds, as we have seen, to the 'topic', which is the old information of a sentence that is known to both the speaker and the hearer. When we say something like 'John saw Bill', then we have been either already talking about John or he is already active in our minds – we might have just seen him walk by, heard a reference to him on the radio, heard his voice, or some such. The principle here is easy: start with the shared information and then give the new information. The new information is provided by the verb and the object together. Each sentence normally tells us something new about something old. So the verb and the object will usually follow the subject. This all means that, no matter what the language, we will generally hear the subject followed by the object then the verb, or vice versa. In other words, the most common orders will be SVO and SOV. In both of these the subject is first and the predicate follows. But this leaves unanswered the question of why

SOV is so much more common. And it fails to tell us why this order is used for gesturing in experiments, no matter what the basic order of the subjects' spoken language.

The advantage of SVO order is that when we talk in a natural and noisy environment, we place the verb between the subject and the object in order to avoid confusing them – that little extra space provided by inserting the verb between them makes it easier to separate the old information from the new in each sentence.

SOV is used even more than SVO, however. There must be something that favors this order even more than SVO. One hypothesis is that objects immediately follow subjects in the majority of languages because humans prefer to present the oldest information – the subject – first and follow that immediately with the newest information, the element that in a sense justifies the utterance of the sentence – what it has to tell us that we didn't already know. That bit of new information is usually presented in the form of a noun and many linguists call the new information the 'focus' of the sentence. This proposal is not original; it is suggested in one form or another by many other linguists. But if it is correct, then what we are really hearing is not the order subject object verb or subject verb object, and so on, but topic focus verb. This hypothesis provides a simple, intuitive answer to a thorny linguistic problem, the order in which we find words in the vast majority of the languages of the world. Of course the terms 'topic' and 'focus' emerge from understanding that the main purpose of language is communication. One should not think of language merely in grammatical terms like subject and object but also in communication-based terms like topic and so on. The meaning and function of these terms is vital to understanding the behavior of real languages. Paradoxically, then, it turns out that SOV order (topic focus verb) is preferred for the same reason that SVO order is also preferred – to communicate clearly in a noisy environment. But instead of keeping track of subject and object by putting the verb in the middle, SOV languages keep track of these items, according to the work of the MIT lab, by adding a suffix to the verb that helps us to distinguish the subject from the object. This allows us to have the topic first, the focus second, and still not confuse them because of the extra 'tracking information' about the subject that is found on the verb (which typically includes such information as whether the subject is masculine or feminine, first person or second person and so on).

This brings us to the observation that we can predict most of the features of the placement of words in human languages by understanding the (communicatively motivated) preference for old information to precede new information, because this facilitates comprehension by adding the new to the old and shared information. Not only is there no need for a language instinct in this explanation, but far and away the simplest and best hypothesis that explains it is that language is a tool.

So could Pirahã speakers be following the same communicative principles as English speakers when they form sentences? Yes. For the most part, the Pirahãs place the topic first, then the focus, followed by the verb. Pirahã sentences are simpler in the number and arrangement of words than English sentences, even though they follow roughly the same principles of organization, because Pirahã sentences are not recursive. But that's not what I want to talk about right now. The striking thing here is that in Pirahã, as in English, form follows function. Our sentences take on the forms that best serve the functions of communication and conformity to cultural values.

The sentences of many languages work like English and so their diagrams will look similar to English's. Still other languages, such as Pirahã, can be dramatically different. Pirahãs often utter forms with meanings similar in complexity to the English sentence 'The people who live in that small yellow house with the picket fence across the street from me raise emus', but they structure such utterances as two independent sentences. Consider an English sentence along the lines of 'The very man who shoots jaguars himself eats pig meat.' In Pirahã it would be:

*Xigihí hi goó kagáíhiai koabáipí. Hi baaí kohoáipí píaii.*
'That very man killed a jaguar. He ate pig meat.'

This Pirahã sequence is like English, except that there is no subordinate sentence. The first sentence has the function of letting us know which man we are talking about, by means of the special word, **goó**, which means that the noun preceding it is in focus, deserving of some special attention. But this is not a subordinate sentence. It is a main sentence, even though it gives us old information. And what follows it is a main sentence, 'He eats pig meat' – giving us the new information about the man we are discussing. Like English, every word of a Pirahã sentence has a function, such as providing new or old information, a meaning, and an order that it must follow relative to the other words of the sentence.

Some theories of linguistics make the foundational assumption that all languages have the same syntactic structure in some abstract way and that the differences we hear are superficial, produced by rules that move things around from the starting order. This movement and the reasons behind it are highly technical and complicated. It is an approach to language followed by a large and influential group of linguists, represented most prominently by Noam Chomsky, and rejected by another equally large and influential group of linguists, represented by Tom Givón and Robert Van Valin, among others.

But as we can see by looking at the Pirahã and English examples, languages that use very different strategies to organize words and information within sentences are equally easy to figure out because the choices on how to order the words are simple: put the word on the left or the right of another, related word. There is more to it than this, of course, but this simple principle will actually get us quite a way in understanding how many languages work. Linguistics can appear harder than it is in some theories. But several people writing these days make the case that human syntax is not nearly as complex as we used to think, in books with titles like *Simpler Syntax* (2005), by Peter Culicover and Ray Jackendoff, or in Robert VanValin's Role and Reference Grammar, as developed in his book *Syntax* (1998). Such studies claim that it is the theories that many linguists employ, rather than the facts of the language themselves, that are responsible for making human grammars look more complicated than they are.

The discussion of Reed-Kellogg diagrams and some basic syntax of English and Pirahã shows us just how simple syntax can be from one perspective, and that grammar is a combination of usually simple forms and simple functions. There are linguistic theories that have emerged over the years based on the same ideas.

If Sullivan's architectural dictum is applied to language, then it becomes as important to understand the functions of languages in different parts of the world as their forms. This would mean that the first task in different languages is to understand what language is for. In other words, what should a properly functioning language do? As we have seen, human languages help us to bring order to our thoughts and to communicate those thoughts to others. Language is for thinking *and* language is for talking. But these functions are only narrowly separated from a practical standpoint. It seems likely that as early humans began to try to communicate with each other they gradually developed a tool for this. And an important side effect of this emerging language tool was that it was useful for organizing and ordering thoughts. The mental housecleaning and thought organization forced by the need to organize words and concepts into different orders, one at a time, for different purposes, would have improved communication and thought simultaneously. More ordered thoughts and communication would have been facilitated in a beneficent cycle of evolutionary improvement.

But language has to have specific properties to channel communication and tidy up our thinking. Some linguists try to identify these properties by asking, 'What knowledge do we possess when we know a language?' The answer to this question has two dimensions: the functional requirements of language – the different meanings it conveys; and its engineering developments – the forms it takes in order to fulfill those functions. To put this in architectural terms, a building may have the function of housing a business, but how well it fulfills that function depends on its design, materials, and construction – its form.

At the center of all this discussion of form and meaning is a deceptively simple idea suggested in the early part of the twentieth century by the Swiss linguist, Ferdinand de Saussure, and further elaborated by a number of others, most prominent among them being the American philosopher Charles Sanders Peirce (pronounced 'Purse'). Saussure's idea was the linguistic 'sign.' According to Saussure, each linguistic sign has a form and a function, which he called either the signifier or the signified. Saussure's idea of signs is not hard to understand. Although street signs are not exactly what he had in mind when he invented the term, a simple stop sign illustrates his idea – that signs are the foundation of all language.

Figure 4 **A common example of a sign**

A stop sign is a red octogan with white letters, on a five-feet-high post, just before an intersection of two roads. That is the form. But what is the meaning of this sign? Roughly it means: 'Halt your vehicle completely just before this sign before proceeding to cross the intersection.' Much cultural knowledge is required to interpret a sign even as simple as this.

Clear enough. And the same connection of form and meaning accounts for words and all other components of human language. We speak in signs. Each word we utter is a sign. 'Brick,' for example, has a form produced by a sequence of English phonemes, /b/, /r/, /I/ and /k/, and a meaning of 'rectangular prism of hardened clay or concrete used for construction.' It is therefore a sign. Human communication almost exclusively involves the transmission and interpretation of signs (interestingly, animal communication in the wild uses no signs). We know that signs are signals transmitted by human bodies. Their immediate source is the human brain, but their ultimate source is a particular cultural history.

Charles Sanders Peirce was one of the most original and brilliant American thinkers of any century. Born on September 27, 1839 in Cambridge,

Massachusetts, he received the first summa cum laude degree in chemistry from Harvard University. Along with his friends Chauncy Wright and William James he became one of the founders of American pragmatism. Peirce should have been a financially successful professor at a major university. Instead, because of his penchant for alienating, rather than cultivating, the rich and the powerful, he lived almost his entire adult life struggling to pay his bills, supported in part by friends like William James, his own writing, and a few other meager sources of income.

In spite of his financial difficulties, though, Peirce wrote prolifically and brilliantly about a wide range of subjects. He has been called the 'American Aristotle.' Bertrand Russell said of Peirce: 'Beyond doubt ... he was one of the most original minds of the later nineteenth century, and certainly the greatest American thinker ever.' Peirce's thoughts on language were among his most perspicacious and influential. In particular, he developed the notion of signs beyond Saussure's original suggestions. For Peirce, the 'sign' subsumes a trio of concepts: icon, index, and symbol.

An icon resembles what it stands for. For example the little pictures across the top of some of the programs you open on your computer are icons each one of which is, usually, a two-dimensional representation of what it does: we see a floppy disk image for the function 'save,' or a picture of a printer for the function 'print.' In natural languages, though, there are sound icons as well as visual icons, including onomatopoeic words like 'woof' for a dog barking or 'whack' for a hit.

Whereas an icon is at some level a representation of what it stands for, an index is not. Rather, it is a physical correlate of something in focus. Dark clouds index rain. Smoke indexes fire. A red, contorted face indexes anger. And so on. Language contains nothing exactly like an index. When we scream in terror or in pain, however, these are indexes, though not part of our language proper.

The most complex sign, however, is the symbol. A symbol associates an *arbitrary* form with a culturally defined understanding or meaning, such as the stop sign above. In language terms, constructions such as 'kick the bucket' or 'the bigger they are the harder they fall' fit here. But all words are symbols. The form of a symbol is arbitrary because its sounds or shapes or colors, etc. are not linked by anything in the real word to their meaning. The links between the form and meaning of a symbols result from cultural decisions. The meaning of the symbol 'run' is associated with the form 'run' in English but it could have any number of other forms. In Portuguese, for example, this meaning is paired with the form '*correr*'. This is only an example, however. Meanings can vary from language to language and different cultures can have symbols that embody meanings not found as such in other cultures.

There is no more fundamental characteristic of human languages than the sign in Saussure's sense, or the symbol in Peirce's. All other components of human language are built on symbols/signs. Signs are socio-cultural contracts entered into by all members of each speech community, in each language of the world. These contracts are underwritten by culture, which provides people with an external, shared memory. Signs are cultural tools and are necessary for us to tell stories and to communicate about a past event, without having to recreate it every time we talk about it. It is crucial to use signs if a species is to attain wishes to achieve human-like communication.

In addition to symbols – the fundamental building blocks of all languages – human cultures also use indexes and icons, underscoring the importance of each of Peirce's subconcepts of signs in human society.

Let us imagine that Tonto and the Lone Ranger are out hunting for a stagecoach robber. 'Where the hell could he have gone, Tonto?' 'Well, Kimosabe,' Tonto begins, 'you see these tracks on the ground? They are the robber's, because there are no others near this burned-out stagecoach and the tracks lead towards Mexico.' 'Tonto, you are one hell of a tracker,' effuses the masked gunman. Tonto shakes his head and texts his wife about this ignorant Texan he is working with and off Tonto and the Lone Ranger ride towards Mexico.

So, what happens cognitively and culturally when Tonto sees the tracks? He recognizes the tracks as an *index* of a horse. He can do this because of his basic ability to recognize foreground and background and because of his cultural knowledge of horses, horseshoes, that Mexico is a perceived safe haven for US fugitives, and so on. He simultaneously recognizes that the index of the horse is also an index of the thief since in his culture the assumption is that the thief is riding the horse – and also since horses leave different tracks, different indexes, when they are carrying the weight of a rider from when they are not. The index is a sensory feature pointing to an object outside of the immediate sensory focus of the individual. Since Tonto *saw* the tracks this means that the tracks are a sensory feature. The tracks point to a horse weighed down by a rider. The index, these tracks, have a physical form and that form points to another object. This is, again, the simplest type of sign. Animals can recognize indexes – dogs track by the index of smell, for example. But there are two other types of sign, symbols and icons. Can animals use these other types of signs?

Recently Diana Reiss of Columbia University and Lori Marino of Emory University discovered that bottle-nosed dolphins are able to recognize themselves in a reflective surface. And that is not all they can do. Reiss and Marion discovered that dolphins look for the best reflective surface in which to view themselves but also that they look longer at themselves when the trainer has marked them in some way that they can see in the mirror. It is almost as though they are admiring themselves or reflecting on their make-up. The dolphins' reflections are signs. But they are less direct than a smell, a track, or some such. They are two-dimensional representations of three-dimensional objects. The dolphins' reflections differ from indexes in that they resemble the things they point to. Horse tracks do not look like horses. But drawings of horses and reflections of horses or actions intended to represent horse walking, horse looks, or some other thing do 'look like' the thing that they are pointing to. Other icons include things like dolls, pictures, carvings, hand gestures, long sentences relaying complex information, and so on.

All humans can recognize icons and indexes. And there seems to be an evolutionary build-up here. If a creature can recognize a symbol, it can also recognize an icon. If it can recognize an icon, it can recognize an index. Although all humans can recognize and use icons, not all societies use icons with equal frequency. In the US, entire businesses have been built around the design and use of icons. Doll manufacturers and movie companies specialize in the marketing of icons.

In Pirahã culture, outside of speech, the use of icons is rarer than it is in US culture. But icons are still important to the Pirahãs. Pirahãs use icons as signs for immediate experiences. For example, when a small plane visits a village, many Pirahã boys and men make balsa wood models of the plane. They paint them, or at least parts of them, with urucum (reddish) plant dye or even their own blood if there are no urucum plants nearby. These iconic planes are used to highlight and celebrate the aviatory visit just past. Many religions, including a number of forms of Christianity, also use icons. But others do not.

In language itself all peoples use iconic signs of some sort – the Pirahãs, English speakers, and others. The Pirahãs speak about little babies in a high-pitched voice, iconic of a baby's own high-pitched voice. Americans talk about French accents with exaggerated 'z's instead of 'th's, as in 'zee man is zee driver.' This is iconic, resembling to some Americans how they think the French pronounce English. The French's own iconic representations of American accents in French include things like heavy accenting of the American 'r' sound and imitating the difficulty Americans have in pronouncing French nasal vowels in words like 'pain' bread.

The Pirahãs have produced an extraordinary set of iconic speech features in order to distinguish the speech of women from the speech of men. There are eight consonants and three vowels in the speech of Pirahã men: the consonants are p, t, k, x (glottal stop), s, h, b, and g and the vowels are i, a, and u. Pirahã women use the same vowels as men, but, with a couple of exceptions, use only the consonants p, t, k, x, h, b, and g. Women do not usually use the 's' sound, so, where men use 's,' most Pirahã women use 'h,' although where men use 'h,' Pirahã women also use 'h.'

This smaller consonant inventory is not all that distinguishes men's *v.* women's speech in Pirahã, however. Pirahã women have a different 'phonetic posture' in their pronunciation. They sound more guttural than the men. They produce this effect by first slightly constricting the walls of their pharynx – the section of the throat just above the larynx – and then by articulating their consonants slightly farther back in their throats and mouths than men do. The net effect of this guttural sound production is that women *choose* to use a smaller anatomical space for talking than men do and they choose to have a smaller set of sounds. The women's smaller set of speech sounds and their smaller vocal space could be taken to be iconic of their smaller physical stature.

In English we use iconic speech frequently. We say things like, 'This is really, really, really big!' or 'I love you SOOOOO much.' In the last example, the extra length and stress on 'so,' like the double repetition of 'really' in the previous one, are iconic signs because their shape (length, stress, number of sounds, and so on) are intended to resemble what they represent – greater intensity of feeling. Resembling is the job of icons. And icons play different roles in different cultures. Examples of this are abundant in religions. Thus, one of the biggest differences in the visual representations of some Christian sects that use icons and Islam is that Islam uses mostly symbols, not icons. Mosques are frequently decorated with symbols, written verses from the Koran, rather than with icons.

But neither indexes nor icons, either separate or together, are enough to result in human language. That is why, although dolphins may be able to use both icons and indexes, they still seem to lack language as complex or as rich as that of humans, even though we know that they can communicate with one another. To make the jump to human language we need Peirce's third type of sign, the symbol. Each symbol is a pairing of a meaning and an arbitrary form to which it bears no resemblance. We could use 'alop' to mean normal-paced bipedal locomotion instead of 'walk', for example. The meaning is either a referent – something pointed to by the word in the real or fictional world – or a sense, such as a quality of an object (color, size, degree of softness, etc.), or an action, and so on. The form can be a sound, a gesture, a series of dots and dashes, a series of 1s and 0s, color-coded plastic shapes, or anything else that humans can come up with. Think about the words 'John' and 'OK.' 'John' *refers* to a human male with the name John. 'OK' does not refer to anything. But it has senses, one of which is 'all is well.' No species other than our own naturally uses symbols to communicate, at least so far as we know. If we ever find spontaneous use of symbols among other species, it will be harder to claim that they lack any sort of language.

To have language, then, humans must first have the cognitive capacity to use symbols/signs. This brings us back to the lack of use of symbols among other species in the wild, although we do see use of indexes and, with dolphins at least, recognition of icons. On the other hand, in specialized circumstances, it does seem possible to train apes and other animals to use symbols. Perhaps the most famous research-er in the training of apes to use symbols is Sue Savage-Rumbaugh, of the Great Ape Trust of Des Moines, Iowa, who has trained bonobos to use symbols. Whether or not every one of her claims can be sup-ported by independent research, the amount of success that she has had that has already been independently verified supports an attenuation of the claim that only humans can use symbols. Thus other creatures are capable of language to some degree. Even though there is little, if any, evidence that apes use symbols in their natural habitats, the fact that they can learn at least some of these shows that the linguistic divide between apes and humans is permeable. And other creatures, such as dogs, seem to have the ability to recognize symbols.

On the other hand, even if humans are not the only ones capable of language at some rudimentary level, we are hands down the best at it. That is because language is a tool *for us* and designed *by us*. And *we set the rules* for the interpretation and use of language. So the deck is stacked in our favor when we ask if any other creature can master language like we do. We should not be too quick, therefore, to draw conclusions about the fact that other animals cannot play the game of language exactly as humans do. The definition of success with regard to language was also designed by humans.

Index, icon, and symbol are all ways of integrating and gleaning meaning from the world. There are no icons possible for words like 'and' or 'to' or 'confabulation.' Therefore, in our discussion of the bio-logical platform for human language, we are going to need to discuss what it is about our bodies and brains that enables us to use symbols.

Questions that all students of language must ask are ones such as: What are symbols used for? Why do we need them? What are the func-tions of our symbol-using? One influential set of proposals on the function and special features of human language was presented in 1960 by the American linguist Charles F. Hockett, in his article 'Origins of Language,' originally published in *Scientific American* magazine. In it Hockett proposed some sixteen features of human language that, he claimed, distinguish human language from all animal communica-tion systems.

Hockett's 'design features of human language,' which are widely accepted, have been discussed a great deal over the past fifty years as researchers have tried to identify what makes human language distinct from other animal communication systems. However, although his observations are interesting, Hockett seems to me occasionally to confuse properties of the mind with properties of language. This confusion is not unique to Hockett. It is seen in various other places in modern linguistics. And yet, such quibbles aside, his observations are so original and such a useful point of departure for discussions of the nature of language that they should not be ignored. Therefore, it is worthwhile to mention just a few of his proposed design features here.

The first design feature is what Hockett refers to as the 'auditory-vocal channel' for communication. While humans can communicate in sign languages, Morse code, smoke signals, and any of a large number of other ways, the normal human communicates by making sounds with their mouths that are heard by ears.

Why is this? Since we are capable of communicating in so many ways, why *do* all humans everywhere communicate mainly vocally?

The simplest answer seems the best one: it is extremely efficient to use the vocal-auditory channel for communication, not because language is designed that way, but because it keeps our hands free. If language were truly 'designed' to use the vocal-auditory channel, sign languages would be impossible. Instead, language is flexible because it has no particular channel as part of its design, even though one channel does work best for most of us. (Sign languages only emerge in special circumstances, such as physical handicap (deafness) or as a lingua franca (a language used for trade), as was the case with the sign language of the American plains Indians in the nineteenth century.) Use of the vocal-auditory channel has freed up our hands for other purposes and is clearly favored in the evolution of *Homo sapiens* – our vocal apparatus shows evidence of a great deal of anatomical change in order to facilitate speech, whereas our hands show no special evolution for gesturing.

Something that has long intrigued me – though I am by no means the only one to have noticed this – is that although humans can communicate well using either the gestural channel (sign languages) or the vocal-auditory channel (speech), they maintain autonomy of form between the two in their grammars. Though all humans use gestures as a crucial component of communication, gestures never mix with vowels and consonants. Have you ever wondered why, for example, making a popping sound by slapping your mouth is not a consonant in anyone's language, a sound of equal status in the language to boring old consonants like, say, 't,' 'k,' 'm,' and so on? Let us use the symbol '§' to indicate mouth-slapping. Then why is there not a word in any language of the world like: *abi§ar*? Or take any hand movement or configuration of American sign language. Why does not any such hand movement ever get used in a word with consonants and vowels in spoken speech?

This is not to say that we do not use our hands when we talk in normal spoken speech. For all of us it is difficult and unusual not to gesticulate while talking. But we do not mix hand gestures with consonants and vowels to produce words, and hand gestures do not function like consonants or vowels in speech. The fact that these types of symbols are never mixed – no words are formed by a combination of movements of hand and mouth in any language in the world – could be explained by the function of language or of our genes. If I believed in extensive a priori knowledge, I might say that we are born with two systems, speech and sign languages, and that we choose one or the other but not both – that they require separate modules of our brain, of our genome, or whatever. The other possibility – the one I prefer – is that speech is more efficient than sign languages, exactly as Hockett said in his original discussion of design features. We can use our hands for carrying things and still speak at the same time. But a mixed system would remove this advantage, since we would have to drop our beer or our firewood or our trousers every time we spoke. It would effectively remove the very advantage that spoken speech has evolved to provide us with. On the other hand, if we are unable to make speech sounds or unable to hear, then sign language is a good fallback position, enabling us to communicate when our evolutionarily principal mode of communication is unavailable. However, it is difficult to talk this way and carry firewood or shoot a bow and arrow at the same time. Therefore, there is evidence even from the main channel that it has adopted that language is a tool.

Hockett also notes that spoken language is characterized by what he calls 'rapid fading.' Because of the nature of sound propagation, almost as soon as we speak the sounds we emit are lost forever. To overcome this problem (not really a design feature, though Hockett calls it that), we have invented writing systems, sound recorders, video cameras, and the like. An advantage of having a written language is that it overcomes rapid fading. It dramatically affects the longevity of language from a few seconds to thousands of years.* Rapid fading limits the use of language, in the absence of writing, to moments of proximity and focus by a speaker and a hearer, a pair of interlocutors working together to construct a mutual understanding.

Some cultures modify their languages to overcome some of the difficulties of the more common channels of communication, gestural and spoken. For example, Pirahã is not limited to these two channels. In addition to being able to speak with consonants and vowels, as we do in English, the Pirahãs take advantage of the fact that their language is tonal – where each vowel in a given word is marked obligatorily with a high pitch or a low pitch (as is also the case in many Asian and African languages). This allows them to whistle their language, which the men do when they hunt; to hum their language, which mothers often do with children; to 'sing' their language (without the need for independent melodies), which all people in the community do to talk about new experiences; and to change all the consonants to 'h' or 'k' and all the vowels to 'a' in order to yell their language across surprisingly large distances, especially across the river (these sounds are partially easier to distinguish, but this speech form is also a result of cultural preferences). Looking at what to us are novel channels of discourse can teach linguists a great deal about the kinds of symbols and communication options that are open to shaping by cultural values.

These Pirahã channels of discourse – yell speech, hum speech, whistle speech, musical speech, and 'normal' consonant and vowel speech – each carry denotational (literal) meaning, connotational meaning (associations with a variety of oneword contexts), and emotional

---

* Also interesting is that in written languages, sentences tend to be longer, more complicated, and, in some writing, less redundant. If this is true, then literacy and writing, which are both cultural inventions, can be said directly to affect grammar. In light of this, it seems unusual that some linguists can still oppose the idea that culture can affect grammar in fundamental ways.

meaning. In all languages, these three different types of meaning result from a confluence of grammar, culture, and cognition. Understanding these different semantic categories is important for understanding how different kinds of symbols and channels can be developed in individual cultures.

Of course, before we begin to talk about the ways in which cultures structure languages, we should be careful to take note of the fact that not all aspects of languages are affected by culture. Languages are structured by cultures and the limitations of human cognition and the pressures of finding solutions to the communication problem common to all humans. We can see this in a simple English word like 'hum.' You can say, 'I can hum the song for you,' 'I can hum the entire song,' or 'I can hum,' depending on your purpose. Some English speakers can quite grammatically say, 'I hummed you the song. What did you think?' But what no English speaker can say is something like, 'I hummed you' (unless 'you' is the name of a song). Why not? Why can't we say, 'I hummed you'? Because English requires that the entity benefiting from an action be accompanied in the sentence by what was produced to its benefit. In a phrase like 'I hummed you the song,' 'you' benefited from 'the song.' Therefore this sentence is fine. But the sentence 'I hummed you' is not OK because the thing you benefited from, 'the song,' is missing. This is the same reason that we can say 'I gave money' or 'I gave money to the church,' or 'I gave the church the money', but not 'I gave the church.' That's not only how English works. It is how most languages work, even those spoken in very different cultures. This construction type may be cultural or the vestige of the cultural influence of the original language of all humans.

Nevertheless, there *is* a way in which culture enters the English language in the verb 'hum.' Surprisingly, not all languages have a word for humming. The Pirahã don't – they just calling it 'speaking low,' but this can also apply to speaking very softly. (The Pirahãs do not whisper, though – probably because tones are lost in whispering.) I discovered this fact about 'hum' while I was completing my master's thesis at the Universidade Estadual de Campinas in Brazil. I was trying to explain the channels of Pirahã discourse to the committee of professors supervising my thesis when I realized that I did not know how to say this in

Portuguese. So I described and demonstrated 'hum speech' and asked my teachers what the word was for 'hum' in Portuguese. They told me that the word I was looking for was '*cantarolar*.' But when they proceeded to describe what it was to '*cantarolar*,' all three of them agreed that it included moving the tongue, so that the action was more like murmuring the melody and lyrics to a song. It is similar to repeating the syllables 'la la la' when producing a melody. (There is no verb in English for going 'la la la,' by the way.) Thus I accidentally discovered that 'hum' in English has a cultural resonance I had not until then appreciated.

Some meanings derive from the human condition. All humans need to distinguish entities from the events they participate in. Linguistically speaking, languages need noun-like words and verb-like words – words that refer to entities and events, respectively or to otherwise distinguish these meanings. We also need to be able to express temporal relations ('John got here before you'), spatial relations ('on top of,' 'under,' 'to the side'), emotions ('Yahoo!'), personal preferences, and so on. No one can give a list of elements that are found in all languages, but we have a good idea of the kinds of things that all humans are likely to express.

Other nuances or types of meaning are specific to certain cultures. In Pirahã there is a part of the verb that tells the hearer whether something was heard about via a third party ('John left –hearsay – the village'), actually observed ('John left – eyewitness – the village'), or inferred from the data ('John left –the evidence suggests – the village').

In English a chocoholic is someone with a compulsive desire to eat chocolate, but there is no equivalent in Piraha. In English language culture, chocolate is a regarded as a readily available treat which in large quantities is bad for you. Words like 'chocoholic' and suffixes meaning 'actually observed' are not universal. It is not that no other culture could talk about these things; It is just that they are significant enough in some cultures to be encoded in special signs.

Symbols derive their meanings from the cultures that employ them. Meaning itself is impoverished without culture. Tarzan undoubtedly had some meanings in his head before he met Jane. But they could not have been very extensive or rich, because depth of meaning comes from culture and society.

English has a large number of constructions that have to do with 'wanting,' 'letting,' and 'making.' There are phrases for 'letting' like 'Let me get back to you on this;' 'Let me talk to him;' 'Let us pray;' 'Let's eat.'

Some semanticists have said that many of the constructions using 'let' in English are not found in any form in other languages, even related ones like Russian and German. At least one researcher has concluded that the greater range and number of 'let' constructions in English compared to other languages results from an 'Anglo cultural preoccupation' – that Anglo adults desire to ensure that their 'wants' are not impeded by other adult members of society.

Such examples show us that we do not need to travel to the Amazon in order to see ways that culture can partially determine the form of symbols and constructions in a given language. For instance, there are studies which show the influence of culture on grammar in Pennsylvania German, especially in the language of Old Order Mennonites; in Northern Iroquoian languages, in Lao, and many others. There are book-length surveys showing the relationship between grammar and culture from a variety of perspectives, using data from various languages and phenomena. Two of the most important books on this topic are *Ethnosyntax: Explorations in Grammar and Culture* (2002), edited by Nick Enfield, and *Deixis, Grammar, and Culture* (1992), a pioneering study by Revere Perkins.

Other examples come from meanings attributed to words that seem at variance with what we would have expected. For example, 'wise man' means the opposite of a 'wise guy.' 'Quicksand' works slowly. A boxing 'ring' is square. 'Stand fast' and 'run fast' are opposites.

Languages are full of these seeming paradoxes. Looking at them from the perspective of culture, however, they make perfect sense. All meanings, literal, colloquial, figurative, and so on, are produced by culture for its own purposes. These words and expressions do not need to match up to their literal meanings. Each one has a history of gradual adaptation from the original functions of its parts to the new functions of the whole. The idea that they are paradoxes seems to derive from a misunderstanding of Plato's notion that there is a fixed, real, and true meaning of a word. But that is not how it works at all. Word and phrase meanings are based on historical accident and cultural preferences.

Television personality Art Linkletter made famous the judgment that 'kids say the darndest things.' He meant that they say unpredictable things, things that are more than merely restatements of what is around them. The human mind is productive and creative, and language has evolved to express this culturally and psychologically driven creativity. But some researchers have confused the creativity of the mind with the creativity of language. Language itself is not creative. It expresses the creativity of our cultures and brains. Ironically, one of the reasons that kids sound more creative in some ways than adults is because cultures also have a 'damping' effect on creativity, reducing it by pressuring members towards conformity as they spend more time in it. As cultural novices, children have not been completely molded.

The late George Carlin, an acerbically acute observer of life and language, had many routines about saying unexpected, probably unique things. He rightly observed, for example, that a sentence like 'I will walk across the room and put Hitler's mother-in-law in my pocket' had probably never been uttered before. On the website Zug.com there is a list of such sentences, including things like 'If you come smoke with me, I'll give you a small bowl of pickles.' Chomsky began this discussion decades ago, in 1957, when he coined his famous sentence 'Colorless green ideas sleep furiously' which, as the wisdom went at the time, has no meaning and yet is perfectly grammatical, showing how grammar not only produces utterances beyond our experience, but utterances that have no relationship to the real world at all. We are able to do this, according to Hockett, because language has the property of productivity – the ability to create new messages by combining already existing signs of the language to form new signs. (On the other hand, in this strict sense, 'Colorless green ideas sleep furiously,' if it truly lacked meaning, would not be a sign, but rather a form without meaning.)

All human languages manifest productivity. But productivity varies in interesting ways across languages, a fact that neither Hockett nor subsequent linguists have explored fully. Productivity can be constrained by culture. The Pirahãs, for example, do not talk about any old thing imaginable, but rather about culturally relevant topics, things that have been directly witnessed by someone alive at the time the event took place. Grammar provides the mechanics of productivity, but that is just one part. So asking how productive language can be is like asking how big an engine can be. Both are limited by the needs and values of the cultures that produce them. Andrew Pawley, a linguist at Australian National University, has shown that in the Papuan language Kalam, verbs describe events in more detail than in many other languages. In this sense, Kalam and Pirahã are similar. Rather than having a simple verb 'to smell', Pirahã describes a situation. I remember the first time I thought I had learned the verb 'to smell' in Pirahã. As I walked past a hut the strong and pleasing smell of freshly grilled fish wafted out of it. I tried to elicit from a passer-by the Pirahã for 'That smells good.' The person I asked responded with a long series of syllables which I studiously transcribed. But I later learned that what I had written down as the Pirahã equivalent of 'That smells good' was 'It pleases me, the smoke, which hits my nose, as I pass by your fire.' That entire sentence in English is a single verb, just one word, in Pirahã. A typical example from Pawley's work on Kalam is:

*am   mab   pu–wk d        ap        agl      kn-la-k*
*'go wood hit–break up get   come   ignite   slept'*
*(third person plural past tense)*

In this example there are several verbs apparently in a single sentence – 'go', 'hit', 'break up', 'get', 'come', 'ignite,' and 'slept,' only the last of which carries the suffixes typical of independent verbs in Kalam. This might sound cumbersome in English, and in fact the literal translation given here is probably not enough for most English speakers to know what the Kalam are talking about, but to a Kalam speaker it makes perfect sense. Pirahã works in a similar way. Such examples show us that literal meaning alone is not enough to understand another language. We also need to know how the native speaker interprets this juxtaposition or string of verbs. They are able to interpret them because their culture filters out unlikely meanings – meanings that do not fit Kalam expectations of what constitutes a likely or normal event.

As Pawley says:

*A fairly literal English translation of (1) would occupy several clauses: 'They went and gathered firewood and brought it, made a fire and slept.' (Mab has the senses 'tree,' 'wood,' 'firewood,' and 'fire.')*

*A free translation might say simply, 'They gathered firewood for the night,' where the act of gathering can, in context, be understood as implying the other activities typically associated with this.*

I have said that languages are made up of signs, Janus-like units where one side is form and the other side is meaning. But there are exceptions. Some parts of language seem to be 'deformed' signs – units where there can be form but no obvious meaning. Or there might even be meanings without forms.

Examples of form without meaning are things like 'cranberry morphs' – parts of words that seem devoid of meaning, as in 'cranberry.' We all know what a berry is. But what on earth is a 'cran'? The answer is just part of the word, a part that contributes to the whole, but which has no individual meaning contribution. We can also see deformed signs in multiple word constuctions, such as in the phrase 'Keep tabs on the dog.' What does 'tabs' mean? It meant something at one time, but in this particular phrase it has lost its meaning for most, if not all, speakers of English.

An example of meaning without form, on the other hand, is when we say something like, 'To run is nice' after seeing a picture of ourselves in the past. We could be asserting that 'running was fun long ago,' that is, we can interpret that 'to run' is referring to the past. But there is no past tense form in 'to run.' Our past tense interpretation comes from outside the forms themselves. The past tense is overtly marked only in other forms of the verb, like 'ran,' 'to have run,' and so on.

We can even see deformed signs in whole sentences. A great example comes from utterances like Chomsky's jabberwocky-like sentence discussed earlier, 'Colorless green ideas sleep furiously,' and its pair he also devised, 'Furiously sleep ideas green colorless.' Neither of these sentences is likely to have ever been uttered before Chomsky constructed them. If grammars were nothing more than a list of statistical regularities, then both of these novel sentences would be of equal status since neither had ever before been uttered. But they are not. The first sentence has an acceptable *form* (adjectives, noun, verb, and adverb in the right order), while the second has neither meaning nor anything like an acceptable form. Chomsky's question is, then: How we can produce a form like 'Colorless ideas sleep furiously' if it does not mean anything? The lesson he presses upon us is that we do this by grammatical rules that are independent of meaning – that grammar and meaning are independent components of our languages. Such examples, however, do not support the idea of an abstract grammar, but rather they buttress Saussure's idea that a linguistic sign is made up of both form and meaning.

It could be, for instance, that although neither of the forms above are signs in English –, since both are devoid of meaning – the first one sounds better than the second because it is statistically more similar to a sign. Computer scientist Fernando Pereira has made an excellent case for this alternative explanation, demonstrating that the word order of the second example is something in the neighborhood of 200,000 times less likely to be encountered than the word order of the first example. Word ordering is subject to statistical generalizations and resulting preferences, even when putting all the words together makes no apparent sense. So while 'Colorless green ideas sleep furiously' does illustrate nicely the distinction between the two sides of a sign, form *v.* meaning, it provides no support for the hypothesis that there is a grammar of the mind that is autonomous from meaning. Simply put, neither of the 'green sleeping ideas' examples belong to English. Neither is a sign. But one of them does bear a stronger resemblance to real English sentences than the other.

However, if we utter the statistically more probable sentence in a normal conversation, a curious thing happens. Native speakers will try to invent a meaning for it. They will do this not just because it resembles an actual English sentence, but also because humans expect all sentences to have meaning. Someone hearing me utter this sentence will assume that, since I have a mind like theirs and since I am a member of their culture, then I must intend for them to understand me. Rather than think that I have uttered nonsense, they will make up a meaning for a sign that otherwise lacks one.

Here is an example of exactly that, though this one did not occur 'in the field.' It is an entry in a 1985 essay contest at Stanford University in which students were challenged to provide meaning to Chomsky's example: The author, C.M. Street, wrote:

> *It can only be the thought of verdure to come, which prompts us in the autumn to buy these dormant white lumps of vegetable matter*

*covered by a brown papery skin, and lovingly to plant them and care for them. It is a marvel to me that under this cover they are labouring unseen at such a rate within to give us the sudden awesome beauty of spring flowering bulbs. While winter reigns the earth reposes but these colourless green ideas sleep furiously.*

This example shows that we humans can attribute meaning to otherwise non-meaning sentences, transforming them into well-formed sentences. But if people try to find meaning in sentences they hear, what does this tell us about the nature of language in its social context *v.* language in thought?

Continuing on with Hockett's design features, another one worth mentioning is specificity. This feature refers to the idea that language has specific functions – communication and thinking – and that it is not used for any other purpose, such as echolocation (the echoes of our speech helping us to find our way in the dark like the sounds emitted by bats and dolphins). But again there is no need for a 'design feature,' because this specificity just follows from the fact that language has evolved to be a tool for communication. Tools have specific purposes. If you want a tool for echolocation, buy a radar or become a bat.

Still, one can see the possibility of other functions for the human voice and language. Walking in the jungle with the Pirahãs, the Banawás, the Paumaris, the Satere-Mawes, and other Amazonian cultures that I have visited over the years, I have found that there is one common sound they all make – a loud, high-pitched yell or hoot, a 'Whoo!' in a falsetto voice. The function of this sound is to communicate to others where someone is or that they have found something worth coming to see and so on. It can be a great way of locating someone who has broken off from the main group of hunters or folks going to work in their jungle gardens. Since these hoots can be used to locate a person or thing, one might jump to the conclusion that they are used in a similar way to the sounds bats and dolphins make to find their way, their echolocation. But no Amazonian group uses hoots like these if there are no other people around – they are not used for echolocation but for communication. They are part of the languages of these peoples and they are, therefore, highly specialized, just as Hockett predicts.

Hockett also discusses a design feature he calls 'discreteness.' Messages in a language are made up of smaller, repeatable parts. 'Run,' for example, is composed of the smaller parts 'r,' 'u,' and 'n,' which are repeated in many English signs. Another part of discreteness, according to Hockett, is that each symbol and each of its parts in natural human languages is perceptually distinct. So the sounds of 'run' – 'r,' 'u,' and 'n' are not like the sounds we hear outside of speech. They are not perceived as blending into surrounding signs but as bounded events that are easily distinguished in principle from other signs around them. The utility of this feature is obvious – we cannot speak in noise, only in controlled, coherent sounds.

Now we come to the most important of all of Hockett's design features, duality of patterning. This feature is what separates human communication from that of most, perhaps all, other species. The way I like to think of this is that language is organized both vertically and horizontally. You can see this in every part of a human utterance. Take an example like 'John kissed Mary.' Horizontally, the sentence is a sequence of three words, running left to right. In normal conversation, to get the interpretation that John is doing the kissing and Mary is being kissed, the words have to come in this order. If you say 'Kissed Mary John' or 'Mary kissed John' or 'John Mary kissed,' you produce either a different interpretation or gibberish or a different stylistic effect, depending on the dialect of English you are speaking.

Some call the positions in this horizontal organization the 'slots' of the sentence (or word, etc). Into each of the three slots of 'John kissed Mary' we can place a different 'filler.' In the first and last positions of the sentence, we place certain kinds of nouns (we cannot, for example, make 'The Blarney Stone' the subject, because rocks cannot kiss). If we replace these nouns with other, appropriate, nouns, we can produce 'John kissed the Blarney Stone' but we cannot say 'The Blarney Stone kissed John' unless we are attributing human-like qualities to the Blarney Stone. The fillers are the 'vertical' organization of human language. Here are some more possible fillers for the slots of a three-word sentence:

| SUBJECT | VERB | OBJECT |
|---------|------|--------|
| John | kissed | Mary |
| Harry |  | Sue |
| The boy | saw | the dog |
| Bill | slapped | the wall. |
|  | believes | that you are stupid |

From a set of three slots and these different fillers, duality of patterning allows us to produce things like 'John saw Mary', 'Bill slapped Sue', 'The boy kissed the dog', and so on. The fillers of a sentence can include, then, words and phrases (such as 'the dog', 'the boy', 'the wall'). But a sentence filler can also include another sentence, as in 'Bill believes *that you are stupid*.' When one filler is of the same type as the structure in which it fills a slot, we refer to the resulting structure and the process that got us there as 'recursive.' That is, the sentence 'that you are stupid' fills a slot in a larger sentence, 'Bill believes …' The property of recursion is an important part of contemporary linguistic understanding of the uniqueness of human language.

But duality of patterning applies not just to sentences; words also show duality of patterning. To see this, we need only separate the slots from the fillers. 'Mary' has four slots: M-a-r-y. Into this we can put different fillers – the phonemes of English. Although the alphabet is the same for any dialect of English, the number of phonemes varies depending on whether you speak Indian English, Australian English, British English, American English, or some other – and even within these dialects there are subdialects such as, in the US, southern English *v.* Boston English, with different numbers of phonemes. We could, therefore, place a 'b' as a filler in the first slot of 'Mary' to produce 'bary' – this is the phonemic form for the name 'Barry.'

It is hard to find clear and uncontroversial examples of duality of patterning outside our own species. But let us say that we can find them – and there is evidence, from the way some animals like birds and dolphins manipulate melodies, that this might be true. This would mean that duality of patterning is a necessary condition for a communication system to be a human language, but that by itself it might not be enough. There is no human language without duality of patterning.

Although examining sets of properties such as Hockett's design features can provide a useful starting point for coming to grips with the complexity of human language, a deeper question arises: Does language have properties that are inherent to it, independent of the problems it aims to help solve (communication and thought) and the resources (cultures and brains) at its disposal?

Although Hockett's design features are useful for describing some of the functionality of human language, there is much more to language than just these features. Languages are intricate formal structures. All signs follow patterns, whether we believe that those patterns are given by rules or statistics. Languages are complex in the organization of their sounds, the composition of their words, and the organization of their phrases and sentences. But even in the light of all other areas of complexity in language, the most complex patterns observed in human languages are not just their phrases and words but their stories, their conversations, their poems. Statistics can predict noun phrases and even sentence structures, but they have a harder time with stories. That is why it makes sense to ask in a village, 'Who tells the best stories around here?' while it makes no sense to ask, 'Who makes the best noun phrases?' Noun phrases are easy. Everybody gets those. Stories are hard. Not everyone can tell them well.

But, as we know, functions require forms. So let us turn to the component parts of language structures. Once we have grasped these, we can ask the more interesting question, which is how or whether culture plays a role in the origin and use of these structures.

One of the components of the complex formal system of a language is its phonology, its sound structure. Sound structure is studied by phonologists, that is, searchers for the truth ('**logos**') about sound ('**phonos**'). Phonology is different from the study of the physical properties of sounds, a science known as phonetics. Where phonetics looks at how sounds are formed in the human vocal apparatus, how they are perceived by human ears, and what their properties are as sound waves travel through the air, phonology is concerned with how people perceive the sounds of their languages and how they organize these sounds into syllables, words, and so on. As examples, take the English sounds 'p', 't', and 'k'. To illustrate that our perception can fool us, depending on the language we speak, hold a thin sheet of notebook paper about one inch from your lips and say the following words at natural volume and speed: 'park,' 'spark,' 'tart,' 'start,' 'core,' and 'score' (phonetically the letter 'c' in these last two examples is a 'k'). What you will see, if you are a native speaker of English, is that the paper moves when you say 'park,' 'tart,' and 'core,' but not when you say 'spark,' 'start,' and 'score.'

This is interesting for a couple of reasons. In particular it demonstrates the complexity of what English speakers have to know in order to be able to know when to produce or not produce 'aspiration,' the little puffs of air that move the paper. First, we have to know that the consonants subject to this rule are those that completely block the flow of air out of the mouth. Such sounds are called 'occlusives.' They are 'voiceless' because when they are produced the vocal chords do not vibrate. Examples of these are 'p,' 't,' and 'k;' examples of voiced consonants, which are *not* subject to the rule of aspiration, are 'b,' 'd,' and 'g.' You can *feel* the difference between 'voiced' *v.* 'voiceless' sounds. For instance, say the sound 's' and hold it for a while. As you produce this sound, place your index finger over your Adam's apple. You shouldn't feel anything. Now make a 'z' sound and hold it. Then place your finger over your Adam's apple again. This time you should feel vibration. That is because 'z' is voiced and the vocal cords must vibrate to produce it. The vocal cords do not vibrate while producing a voiceless sound like 's.'

In order to be able to speak in English, you need to know more than how to make the sounds and which sounds to apply the rule of aspiration too, though. You also need to know that the aspiration rule only happens at the beginning of syllables. There is no puff of air emitted with the pronunciation of voiceless occlusives in words like 'score,' 'spark,' and 'start,' because in these words the occlusive consonants are not at the beginning of a syllable (the 's' is in each of the example words). Who taught you syllable structure? No one. If you are a native speaker, when did you know the syllable structure of English? Before you started kindergarten. This kind of complexity and how humans come to master it is one of the reasons that phonology is so interesting.

Not only do you need to know syllable structure and the difference between voiced and voiceless consonants to produce aspiration correctly in English, you also need to know whether or not the syllable containing the voiceless occlusive – also known as a voiceless 'stop' – is stressed or not. Stress refers, to put it a bit simplistically, to greater energy being used on a particular syllable, usually in the form of greater loudness, higher pitch or more length of the stressed syllable. This is why even the 'p' at the beginning of a syllable in a word like 'open' is not aspirated – because the stressed syllable in this word is 'o,' not 'pen.' If we shifted the stress to the syllable 'pen,' however, then the 'p' would be aspirated. Try pronouncing 'o-pen' v. 'o-pen' with a sheet of paper in front of your mouth and you will see a puff of air come out in the second example, but not in the first.

This example illustrates not only how much we know about our language as native speakers, but how we are able to manipulate our language through that knowledge to fit a range of situations, from serious talking to joking. It is striking to observe how humans routinely use and repackage their own language in ways so subtle that without training we could not begin to explain how we do it, nor could we teach someone else how to do it. We know all about aspiration, syllable structure, stress, and so on. We just don't know that we know, until a linguist shows us – we seem like the servant boy talking to Socrates. Only seem, however. Given the evolution partnership between our ears and vocal apparatus such distinctions are easily learned.

Although words are the building blocks of everything else in grammar, it can be quite challenging to convince the beginner linguist that they even exist. Yet this challenge largely results from skepticism, rather than lack of evidence.

For most people, the main reason why proving the existence of microbes is so difficult, for example, is that they do not have the microscopes with which to see them. One of the most difficult parts of learning a foreign language is learning to recognize the words in a stream of speech. In the normal rate of speech of any language, there are rarely pauses between words, just a current of sounds. If you are a fluent speaker you know what to listen out for; if you are not, you do not. This is why so many people learning a new language say of the speakers of that language, 'They talk too fast!' There are some differences in speech rate between speakers, but not so great that we could say with any accuracy that 'Spanish speakers speak faster than English speakers' etc. It is not that they talk too quickly, it is that the learner listens too slowly, because the learner is listening in ignorance. Examples of this are abundant in all languages, even English.

Let us say that it is about noon and you are hungry. You speak southern Californian English. So you turn to your friend or spouse or class and you say 'Squeat!' Everyone understands you, except the occasional foreigner in your midst. What the hell is 'Squeat'? Said more slowly it becomes 'Let's go eat.' Say it as fast as we normally do in American English and it comes out as 'Squeat,' even though we hear it differently, as 'Let's go eat,' which is itself a reduced form of 'Let us go and eat.' Sounds like one word but it is really four.

There are kids' jokes that rely on phrases sounding like words, as in: 'When you're dancing with your honey and her nose is kinda funny, well, you may think it's funny but it's not' – the last two words of which, when said quickly, come out as 'it's snot.' There is another one that does not work for me, because my pronunciation is different: 'If you were stranded in the desert what would you eat? The sand which is there.' Spoken quickly the last line, for those who do not pronounce the 'h' of 'which' as I do, comes out as 'the sandwich is there.' Another puerile giggle.

Whether written or spoken, and however mismatched the form and the meanings might be in actual speech, all languages depend, to a greater or lesser degree, upon a set of shared structures. Metaphorically, the architecture of language looks like the letter 'T.' Some refer to this conception as the 'T-model,' which is one way to conceptualize what happens when we form a sentence. This model is represented in the following:

### Figure 5 **T-Model of Language**

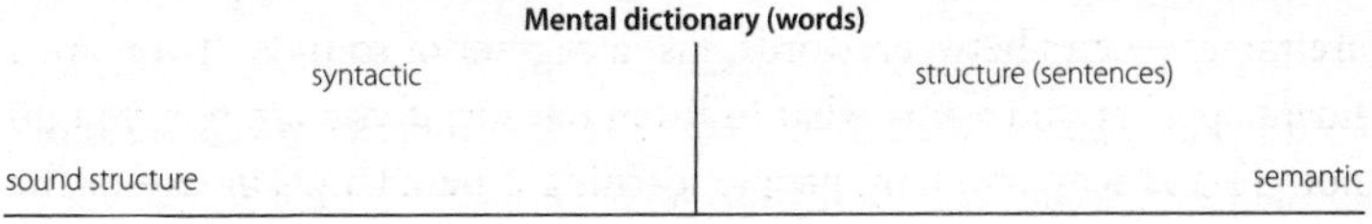

| | Mental dictionary (words) | |
|---|---|---|
| syntactic | | structure (sentences) |
| sound structure | | semantic |

Words are the starting point of the T-grammar because everything else in a grammar is built on words. Words are kept inside our mental dictionary, somewhere inside our long-term memories, organized in various ways, such as by their meanings, by their frequencies, and by the order in which we encountered them.

When you utter a sentence, you begin by picking your words – probably leaving some of the word choices until more of the sentence is formed, moving with amazing speed from the unconscious decision to put some words first until your lungs stop expelling air at the end of your utterance.

Morphology is the set of principles that regulate the way a language forms words. The brain must be able to encode (assemble) and decode words. Morphology is in a sense the theory of how our brains do that. The components of the structure of a word can include things like number (whether the subject is singular or plural, for example), person (whether, say, the subject is 'I,' 'you,' 'she,' 'he,' and so on), or whether the word indicates possession (as in 'John's book'). Different languages form words in different ways, but in all of them the markings on words and arrangements of syntax that are seen on words and sentences have one simple function: to help the brain decode the message and understand the signs in the transmission with greater facility.

There are different ways to reduce (or increase) complexity in languages, however. Complexity is largely a function of particular cultures. One simple example is the effect of writing on a language. Once a language is written down it almost always develops two styles, one written and one spoken. These two styles contrast in complexity in different contexts. Researchers have demonstrated that in some circumstances the written syntax is more complex than the spoken syntax, such as long sentences in novels, while in other circumstances, such as queries to databases, it has been shown that spoken forms are sometimes longer than written forms.

The T-model begins with the assumption that all humans carry in their heads a mental dictionary of the words of their language. A Pirahã speaker, for example, not only knows the concept 'wooly monkey,' but also has in their brain a memory that the word for this is **kapóogo**. A great deal of research over the years has shown that this mental lexicon is more than a list of words. People organize their words by related meanings (semantic fields), by sound structure, by most common meanings, and so on. Words are accessible to our memories from a variety of different directions. A crucial part of our storage of words is the appropriate contexts in which each word can appear. Any Pirahã knows that **kapóogo** can be used as the object of a verb like 'to eat,' but that it cannot be the subject of a verb like 'to fly.' They also know how to say the word, what other words sound like it, what other words are similar to it in meaning, that its meaning is more closely related to words for other animals than to those for plants or minerals, and other facts that indicate long-term storage of information about wooly monkeys.

Figuring out what the words of a given language are can be very hard work. Before a scientific linguist reports about words – or any part of the grammar – in some language, they often demonstrate that the language in question in fact has words. It is not that anyone doubts that there are words in all languages. But linguists require that the evidence be marshaled in support of the existence of words, language by language. And it must further be shown that the words of a language have the forms and meanings that the linguist claims they do. This is not as easy as it sounds. Think about what you would need to do to prove that English, say, has words and you will better understand what words are and why they exist. I used to ask my students after we had discussed the nature of words and their structures if they could prove to someone who had never had such a class that words actually exist. Some would reply, 'That's easy. Words are the things between spaces on the written page.' But you cannot define words according to the writing conventions of a particular language, because all languages, even unwritten languages, have words. A definition of words based on printing will not apply to an unwritten tongue. This graphic definition of words will not even work with all written languages, because some written languages – ancient Greek being one example – do not put spaces between words. To the ancient Greeks the sentence I just wrote would be written like this:

*somewrittenlanguagesancientgreekbeingoneexampledontputspac-esbetweenwords*

Let us say that a speaker of some other language told you that 'squeat' was one word in American English. If you agree with me that 'squeat' is simply a fast pronunciation of three words and not one, how would you convince them that you are right? This is a common linguistic problem. And linguists use several different lines of argument here. But the main reasoning is based on the concept of distribution – you can recognize something as an 'x' if it is found only where things like it are found. It and the rest that are found there can be called 'x's. Everything else will *not* be an 'x.'

Take a stream of speech coming out of someone's mouth and write down all the sounds – something you can do for any language on earth if you have mastered the International Phonetic Alphabet (IPA). In this stream certain kinds of sounds are repeatedly grouped together in various positions. Let us say you hear a short little stream like 'Leonard loves Billie Jean.' You write it down in your notebook. Then you hear another person say, 'Yes, and Billie Jean loves Leonard.' Even if you do not understand a word of English, you can see that the sequence of sounds (ignoring the IPA for the moment and just using the English spellings), 1-e-o-n-a-r-d and b-i-1 -1 -i-e-j-e-a-n occur in different positions in the stream. You might then write down other sentences like 'Leonard saw Billie Jean,' 'Leonard loves bacon,' 'Billie Jean loves potatoes,' 'Billie Jean eats bacon,' and so on. If you keep working at this long enough, you will eventually notice that the sequence of letters that represents events or states can change in ways that sequences like 'Leonard,' 'bacon,' and 'potatoes' do not. So you will be able to see 'love,' 'loved,' 'see,' and 'saw,' on the one hand, but not 'Leonarded' or 'Bawlie Jean' on the other.

To keep life simple you would probably want to generalize the lessons about sequences you hear. So maybe you want to say that sequences like 'bacon,' 'Leonard,' and 'potatoes' are 'names for things,' or, to use a more economical term, 'nouns.' Words like 'love,' 'eat,' 'see,' and so on you could call, let us see, 'verbs.'

Continuing to use distribution – repeated patterns of sounds in different locations – in this way, you can build up classifications of nouns, verbs, adverbs, adjectives, and so on, as well as subclassifications of each of these, such as 'proper nouns' ('Leonard,' 'Billie Jean') *v.* common nouns ('bacon,' 'potatoes'). To put all of these in a single class such as 'the class of things that sentences are made of' you could call them, wait, uh … 'words.' The evidence for words is more complicated than this ultimately, but it is more or less how linguists working on languages that no one has ever studied can analyze their words.

But what about 'squeat?' How can you tell that 'squeat' is four words and not one? Well, if you observe carefully, distribution gives the answer here, too. We cannot say, 'Leonard loves squeat.' Nor can we say, 'Billie Jean saw three squeats' or 'squeats run fast.' But we can say, 'I said squeat.' 'Squeat' appears only in the positions where you could also get the phrases 'Let us go eat,' 'Let us jump in the lake,' and so on. That is, it has the distribution of a certain kind of sentence (an 'exhortative' if you must know). When we say, 'Let us go eat,' we establish a clear correspondence between the sounds or phonology of words and their meanings. One word, one primary meaning. But as soon as we begin saying things like 'Let's go eat,' with the 'us' reduced to 's,' we confuse the sound–meaning relationship – a mismatch that is particularly acute with 'squeat.' Four primary word meanings reduced, apparently, to only one word. This makes it harder to recover what the signs are.

Contractions like 'let's' and 'squeat' work only because native speakers can understand other native speakers with less perfect correspondence between form and meaning than a non-native speaker needs. That is because native speakers know what the likely sequences of words are in their language. They also hear themselves produce contracted forms like 'squeat' and they know both what meaning they intend by such forms and what the full forms of the words are that underlie them. This has a very interesting result – native speakers will hear things that are not there. A native speaker will usually hear 'squeat' as 'Let's go eat,' not as a monosyllable, except when it occurs out of context or in the focus of discussion as it is here. A non-native speaker, however, only hears what is there. They lack the native speaker's knowledge of the signs of the language. The learning of a foreign language entails the learning of signs – grammar, phonology, and semantics – *and* the ways in which the relationships between them can be skewed. Here is the basic relationship for 'let us go eat,' taught perhaps to learners of English:

| Let | us | go | eat. |
|---|---|---|---|
| *Expression of desire for permission or agreement* | *speaker and hearers* | *remove to a place consume food.* | |

In the slow form of the sign, a sentence, the correspondence between the individual smaller signs, the words, is clear. The faster we say this, though, the fewer the sound clues the listener has to link form to meaning:

**Squeat**

| *Expression of desire for permission or agreement* | *speaker and hearers* | *remove to a place consume food.* |
|---|---|---|

Once again, this is why many people think that the speakers of a language they are learning talk too fast. We *all* talk just fast enough to be understood and to convey the emotions we are feeling to our fellow humans. Language is a tool.

English morphology is not all that challenging. So English speakers might not think morphology is very hard. English verbs, for example, have a maximum of five distinct forms: 'sing,' 'sang,' 'sung,' 'sings,' 'singing'; 'hit,' 'hits,' 'hitting;' 'eat,' 'ate,' 'eaten,' 'eats,' 'eating,' and so on. To see the meanings of these distinct forms, consider the verb 'to sing.' 'Sing' is the form used for the present tense for all subjects except third person singular: 'I sing,' 'you sing,' 'we sing,' 'they sing,' but 'he/she sings.' 'Sang' is the simple past tense for all subjects: 'I sang,' 'you sang,' 'he sang,' etc. 'Sung' is the perfect form, meaning a past event that had relevance to some past or future event, as in 'By the time I arrived, she had already *sung*' or 'By the time I will have arrived, she will have *sung*.' 'Singing' is the form used when an action takes place over an extended period of time, as in 'He kept on singing far longer than I wanted!' Or 'singing' can also be used to make the action a kind of thing, taking the form of a noun, as in 'singing makes me happy' – in this form *sing + ing* together have the distribution of a noun (they are only found where we find nouns), whereas the other forms of 'sing' are found only where we would find verbs.

Verbs like 'hit' are even easier: 'I hit you yesterday;' 'I will hit you tomorrow;' 'I will have hit you;' 'I had already hit you;' 'I was hitting you; "hitting can be a pleasant activity when you are the one hitting;' 'he hits you.' Only three forms of the verb for all that meaning!

But in other languages, words can be much more complicated. In Portuguese, for example, like some other Romance languages, each verb can have from thirty to fifty forms, depending on how you count them. This larger number of forms does not mean that Portuguese can make more distinctions of meaning than English, however. What Portuguese does with word forms, English does with series of words. For example, in Portuguese you can say: 'estou' 'I am;' 'estava' 'I was being' or 'I used to be;' 'estive' 'I was.' Two lessons of general importance can be learned from this simple list. First, English can make all the distinctions Portuguese can, just not with individual words (otherwise the translation would be impossible). Second, English's phrasal forms (technically referred to as periphrastic forms because they use multiple words to get across what Portuguese can do, more or less, with a single word) can express some information less ambiguously than can be done in Portuguese. Take, for example, 'I was being' v. 'I used to be.' If I say in Portuguese 'Estava no mercado,' this can mean 'I used to be in the market (but I no longer am)' or 'I was in the market for a period of time.' The first translation is the English habitual 'used to' and the second is the English progressive 'was …-ing.' The Portuguese forms like estava are therefore ambiguous, they have variable meanings, but the English forms are unambiguous, they have a less variable meaning.

So when we say in English 'I used to eat breakfast every morning,' this means two things: 'it was my habit to eat breakfast in the morning' and 'I no longer do so.' But in Portuguese 'Tomava café de manhã' can mean either 'I was having breakfast in the morning' or 'I used to have breakfast in the morning.' (It literally means 'I was drinking coffee' or 'I used to drink coffee,' but this expression idiomatically means to eat breakfast – a culturally affected meaning, since the common Brazilian or Portuguese breakfast is mainly coffee with some bread.) The lesson here is that, just because one language has more forms of words than another, does not necessarily mean that you can say more things in it than in the language with the simpler morphology.

The Pirahã language, like many other American Indian languages, has a word structure or morphology so complex that the morphologies of Portuguese and all other European languages seem simple by comparison. A Pirahã verb, for instance, is composed of multiple verbs put together and a series of as many as sixteen suffixes. The suffixes include things that English and Portuguese speakers would find unusual like '-ábai,' 'frustration of an event near its end' and '-ábagaí,' 'frustration of an event at its beginning.' To illustrate these, because they are so foreign to us, consider this situation: a Pirahã man pulls back his bow; just before he releases his arrow, the bow string breaks. He would say, 'I shot-*ábagaí* it,' meaning 'I was about to shoot it but something prevented me from shooting.' Or if he shot but missed he would say, 'I shot-**ábai** it.' A philosopher friend once pointed out to me that these suffixes would have interesting applications in describing sexual performance, too. An interesting idea – but I haven't yet checked it out.

Pirahã also distinguishes the suffixes -**b** *v.* -**p**, where the -**b** means 'action done in a downward motion' and -**p** means 'action done in an upward motion.'

At the opposite extreme from Pirahã and other American Indian languages, there is Mandarin and some other Asian languages that lack almost all morphology – their words do not vary even as much as the five forms of English. Variations on meanings are expressed periphrastically, with multiple words similar to phrases like 'used to eat' in English being used widely.

Putting a sentence together, on the other hand, is a different story. Sentence formation entails, according to some conceptions, that words are pulled from the dictionary and then inserted into a template (a mold or form) for phrases of their language. There are a limited number of possible permutations of word orders in the world's languages. This means that one could model the different word orders based on a set of rules, Chomsky's traditional approach, or, alternatively, list templates of word structures or sequences, and select among them according to their statistical preference in a given context, as many computer scientists and non-Chomskyan linguists have suggested. Culture could favor some templates over others. Robert van Valin of Düsseldorf University has developed an entire theory of grammar that shows how simple the templates for particular languages can be.

After we have put our words into the final form of the phrase we are uttering, we pronounce our sentence. As we do this, almost instantaneously, our hearer decodes what we have said. The speed of this operation is astounding. We might expect that another native speaker of English would need to stop and think about what we are saying before understanding it. But, no, the understanding seems to be immediate. And the response is often on its way before the speaker has even finished their initial sentence. For this decoding to work, the hearer has to be able to immediately analyze the sounds coming through the air to their ears and assign meanings to sets of sounds (words and phrases). In fact, as we saw with Chomsky's famous example 'Colorless green ideas sleep furiously,' the hearer will work very hard to attribute a meaning, however far-fetched, to any sentence they hear. This is because the hearer believes that the speaker is trying to communicate and, since we all know that meaning is essential to communication, we stretch and strain to figure out what the meaning is. If there isn't one, we will often go so far as to try to invent our own. This seems reminiscent of our relationship to art. In fact, for most things that we do not understand at first, we try to find a meaning.

The analysis of sentences and their parts is only a small part of the understanding of the grammatical structures of a particular language. All sentences are part of a hierarchy that begins with conversations and moves down to stories (stories are part of exchanges or conversations between humans), ultimately ending in words and their parts. For this reason, the most important kind of data that I collect as a field researcher is the 'discourse' – stories. Stories give us language data in a natural context. Stories are crucial to any understanding of the nature of language.

One way that I work with stories in my field research is first to ask one speaker to talk about some topic, saying whatever he wants to say. He might tell me about his most recent fishing trip, his child's illness, about going to his jungle garden, and so on. When he leaves, I bring in another speaker. I replay what the first speaker said to the second speaker and ask him to tell me what the first speaker said. This never results in complete repetition. What it gives me in fact are two versions of the same story. Then I might ask the second speaker why he said it one way while the first speaker said it another. Sometimes the second speaker says, 'They are the same,' meaning that there is no significant difference in his mind between what he and the first speaker said. Other times, however, the second speaker will say things like, 'Ha, ha. He spoke ugly. We do not say that. We say this.' In this way the second speaker acts like an editor for the first speaker, appealing to some unwritten standard of grammatical correctness that underlies this unwritten language. But neither speaker, just as no untrained speaker of English or any other language, can explain *why* the first speaker is wrong, except in impressionistic, unscientific terms. And yet the stimulus of hearing another person's 'dead speech,' not coming at them in a live exchange but objectified on a recording, elicits a flood of opinion that must be based on tacit knowledge of their grammar. Grammars inhere in languages. Languages are built around the principles determining how stories and sentences are put together. Somehow every speaker comes to have a knowledge or opinion of what sounds right.

But both at the story level and at lower levels of the hierarchy of grammar out of which languages are constructed, it is intriguing that native speakers 'bring interpretation' to the task, so that what a native speaker derives from utterances is a gestalt result – that is, the whole is greater than the sum of its parts. This is seen in the inability of native speakers to hear differences in their languages that outsiders perceive easily. At the level of conversations we see this whenever one person overinterprets someone else – where they 'read between the lines' of what another native speaker has said to find a meaning that is not literally present. 'I am not saying you cannot go,' for example, said by a wife to a husband heading out for a beer with his buddies, might be interpreted by the husband to mean exactly that he should not go, even though the literal meanings of the words do not say this.

At a lower level, let us consider English aspiration examples. Why can the average speaker of English not hear the aspiration of voiceless stops at the beginning of stressed syllables? After all, many people learning English for the first time often hear the difference between aspirated *v.* non-aspirated consonants. They do not know what it is and it may confuse them, but they do hear it. Yet to native speakers the aspirated consonant sounds identical to the non-aspirated version. To us, in normal speaking, an aspirated 'p', written phonetically as **p$^h$**, sounds just the same as the unaspirated **p**. They sound the same, that is, until we put each of them in the other's place. Pronounce the word 'spark' with an aspirated 'p' and it will sound very strange. In fact, it will be hard for you to do this. Your knowledge as a native speaker will make you resist aspirating that 'p' sound. So hold up the piece of paper to your mouth again and make it move with a puff of air as you say 'sp$^h$ark'. Now can you hear the difference? It's strange, isn't it? Why is there such a difference in perception?

Although the sounds of our languages are real, physical entities, our perception of them is based on our expectations. As Paul Simon once sang, 'a man hears what he wants to hear and disregards the rest.' We often hear what is not there and do not hear what is there. Aspiration is an example of the latter illusion. In fact we could say that languages are built on sound illusions, as we saw with 'squeat'. These design features are important in trying to understand the nature and uniqueness of human language. But they seem less like design features of language than general properties of the human mind. They can each be interpreted as facets of human cognition, rather than of human grammars or communication.

Take Hockett's 'learnability' feature, for example. We humans can certainly learn languages. But that is like saying that air is breathable. If air were not breathable, we would be dead. The breathability of air is not a feature of air but a feature of life as it has evolved on this planet. More appropriately, since we are talking about cultural artifacts here, saying a language is learnable is like saying a car is drivable. That is what cars are for. A car that is not drivable will not be adopted by any culture; likewise a language that is not learnable.

Or consider his feature of 'productivity.' This is, again, just language doing its job. If I want to talk to you about any experience I have, then my language needs to be designed to do that. For example, the Pirahãs do *not* talk about everything. Not even in principle. Because their culture constrains the subjects that they can talk about. Neither do the Kalam of Papua New Guinea. Nor do Americans. Nor the French. And so on. Even our verb structures, from those of Kalam and Pirahã to English, are constrained by our cultures and what these cultures consider to be an effable 'event.' The Pirahãs, for instance, do not talk about the distant past or the far-off future because a cultural value of theirs is to talk only about the present or the short-term past or future that are either typical activities for the Pirahãs or ones which have been witnessed.

Hockett's design features are a major step forward in bringing linguistics into the sciences from the humanities – which for some, but by no means all, researchers indicates that linguistics has 'come of age.' But most of these features are simply about the physics and use of language, rather than properties unique to linguistic entities *per se*. And they have nothing deep to tell us about meaning. Yet since meaning underlies communication systems, we cannot talk of language without it. It is essential to understand the meaning of meaning better. Only then will anyone be in a position to fully understand the development and nature of human language.

Meaning is all around us. Cultures provide it and we must all learn to find it. For example, friends of mine emerging from the theater after watching the pathbreaking movie *2001: A Space Odyssey* in the 1960s said things like, 'It is visually stunning. But what does it mean?' It is a sign of intelligence and *savoir faire* to have something to say about the meaning of *our* cultural outputs like movies, songs, and novels. And our quest for meaning goes beyond words. Perhaps the biggest question of our existence is 'What is the meaning of life?'

Meaning is what our speaking and listening are all about. Meaning is not found merely in what is said, but in what is heard and how the information that is heard is responded to. It lies in hearers responding to the forms of their language, its words, sentences, and stories.

Meaning in individual languages evolves over time through the history of interactions of a group of people in a specific culture. Culture creates or mediates the entry of meanings into languages. As an example, consider a westerner walking along a jungle path with an Amazonian Indian. Both hear a thump. The westerner wonders what it is. The Amazonian knows that the thump comes from the sound of a

Brazil nut pod hitting the ground after falling from the tree. He tells the westerner that this thump 'means' that the Brazil nuts are ripe.

Or suppose that these companions look into the sky and see the clouds gather and darken. The Amazonian says to the westerner, 'This means it is going to rain.' Likewise, when an Amazonian Indian is with me in a city for the first time, I need to explain things like 'When this light turns red [literally "looks like blood" in Pirahã], it means that we cannot go forward.'

Or imagine that someone addresses you in a foreign tongue and you ask them, 'What does that mean?' or 'What does that mean in your language?'

Meaning is the map of our words to their concepts or to the things our words stand for or represent in the non-linguistic world. In connecting our words to the world, meaning is the roadmap of our existence.

Not surprisingly, meaning is not simple. Who gets to decide which meanings are right? Noah Webster? The Oxford English Dictionary? The frustrating answer is that the meaning you attribute to a word or utterance is partially up to you and partially up to whoever you are communicating with at the time.

The ancestors of *Homo sapiens* developed proto-cultures and proto-languages – precursors to modern languages and cultures. It would be wonderful to be able to travel back in time to the African savannah hundreds of thousands of years ago and watch the social and linguistic development of meaning among early hominids. If I were able to make such a journey I would like to investigate two questions in particular. The first is: 'What did these pre-human apes communicate about?' The second is: 'How did they transmit messages?' They could not have talked like modern humans because they lacked the vocal apparatus to do so. Fossil analysis has shown that the anatomy of their mouths and throats was very unlike that of modern humans and so they could not have produced speech like we do, any more than a modern chimpanzee can. On the other hand, pre-humans must have communicated with one another prior to the evolution of the modern human vocal structure. It is hard to believe that humans waited until their vocal abilities were improved through hundreds of thousands of years of evolution before they thought to put this mechanism to work. Evolution is a tinkering process, in which biology responds gradually to a creature's needs by producing random genetic changes that count as improvements in some circumstances and death sentences in others. If these hominids were already trying to communicate orally, then the physical differences between their vocal structures and ours – we have increased length between larynx and mouth, shifting the tongue back into the upper part of the throat relative to other primates, and so on – make perfect sense, because such changes would have enabled us to become more effective speakers, projecting audible and distinguishable sounds across much greater distances. Natural selection exerted pressure to favor effective communicators over less effective communicators (or we wouldn't have language).

Without such a vocal tract *Homo sapiens* would have an anemic communication system. On the other hand, there is definitely a sense in which information could be received and action be taken based on other types of communication systems. Some philosophers have used similar reasoning to argue that thermostats are intelligent – the switch on the air-conditioning system understands and responds to information (temperature) from the environment. But maybe thermostats are not communicating. Maybe they are no different from a rock that is heated until it explodes. Even I agree that we should not describe this by saying that the rock received information – 'It's getting hotter' – from its environment and then put this information to use by exploding. The same thing might be said about the hypothesis that trees communicate: they take in information from the length of the day and use this to 'decide' to drop their leaves. This can be characterized as changing behavior (holding leaves to dropping leaves) via exposure to new information (shorter days). On the other hand, as with thermostats, so perhaps with trees, the alternative explanation might not be about information but about mechanics – maybe shedding and growing leaves is not communication, but merely physics – it depends on whether you believe that sentience is required for communication or merely information flow.

It is not easy to tell what communication is in some cases. Consider ant communication. Do the *formicidae* intentionally communicate via pheromones or do they drop pheromones involuntarily, like trees drop leaves, and do other ants respond in equally physical, non-communicative ways? If emission and responses to pheromones are involuntary, then maybe ants do not communicate either, or not with any intentionality at least.

Whatever the case turns out to be for trees, ants, paramecia, or other members of the kingdom of life, communication is certainly not the unique province of *Homo sapiens* – we see that unambiguously in most mammals and in many if not all birds. The human case is also almost certainly related to what we see in other creatures.

What this all seems to show us is that an examination of the structures of language, its sound structures, word structures, sentence structures, stories, and so on reveals that linguistic structure is like architecture – form follows function. Our language is shaped to facilitate communication. There is very little evidence for arbitrariness in the design of grammars. Each language will, of course, through the accretion of the centuries, have arbitrary forms that simply have to be learned and which have no obvious explanation in terms of communication or current cultural values. They are just there. But once we begin to take seriously the idea that culture is a separate force from the need to communicate effectively and that these two forces work in tandem, along with the limitations and design features of general human intelligence, then grammar, phonology, and morphology all make more sense and become less mysterious. Language ability begins to look less special and more like other cognitive skills.

We also see that people both overinterpret and underinterpret what they hear based on cultural expectations built into their communicative patterns – Kalam speakers are able to understand strings of verbs that are quite unlike what we are used to in English. English speakers do not hear aspiration or its absence, yet this distinction is vital for speaking English intelligibly. Culture, cognition, and communication are the shaping forces of our languages and, though all are necessary, none alone is sufficient to produce language.

## Chapter Seven

# THE PLATFORMS FOR LANGUAGE

*'The question is not whether anything at all is specific to human beings and/or hard wired into the brain, but whether there exist rules that are specific to human language and not a result of our general conceptual/perceptual apparatus together with our experience of the world.'*

Adele E. Goldberg, 'The Nature of Generalization in Language' (2009)

Many years ago the Brazilian-Israeli philosopher Marcelo Dascal and his Uruguaian-Israeli wife Varda joined me among the Pirahãs for about a week. While there, they asked the Pirahãs questions about animals, trying to get to grips with the Amazonians' philosophy of animals and how they relate to humans. One question was whether monkeys could talk or think. Varda in particular emphasized to the Pirahãs that, to her at least, monkeys look like people. The Pirahãs responded that of course monkeys cannot talk. And to the Pirahãs, they don't even look like people. When you get right down to it, they said, monkeys look like dogs – they are all covered in hair. The Pirahãs believe that only people can talk. The principal reasons for their belief are biology – people are people and other things are not, period – and culture – people talk differently because of what they eat and how they live. So the Pirahãs recognize that language is something only humans do – it is part of human biology and culture. Most of us would agree with them. And all of us would agree with linguist Noam Chomsky when he says that his human granddaughter is better at language than her kitten. But such obvious observations do not get us to a very clear understanding of what our biologies and cultures must be like for us to have language. Nor do they help us understand the necessary linguistic components of any human language – the things without which no language could exist.

Language is delightfully complex. There are so many things going on whenever we speak or understand things spoken to us that I have often wondered how language is even possible. How is *Homo sapiens* able to have a language, a communication system so marvelous?

This is different from the question that began the previous chapter, 'How do you build a language?' That question asked what the basic components of human language are. But the crucial issue here about the language platform is what human bodies and brains need to be like for us to have and use language.

A primary building block would be the ability of humans to invent words. Without words there is no language. There is clearly something about humans that make it possible for us to create words, and that something is not found in other species. We have the capacity to join meanings with forms to produce word signs, whether the forms are sounds or gestures. And there is also something about humans that enables us to put words together to make whole thoughts and sentences.

Biology is fundamental to give form to speech. Language can take many forms compatible with human biology – alphabetic writing, Braille script, sign languages, speech, and so on. The most common form for modern humans is speech – using consonants and vowels pushed through our anatomically modern vocal apparatus. And yet we were not always capable of this. If our pre-*Homo sapiens* ancestors used speech before their voice box, larynx, and the shape of their entire vocal system had developed into its present shape – which is radically different from the same apparatus in other apes – then they could not have made many sounds. Hominids' entire throat, tongue, teeth, and mouth structure had to change before they could physically accomplish that. The first speaking hominids might have chatted away merrily while sounding like their mouths were full of marbles (as research on pre-*sapiens* speech has suggested).

Human tools – of any kind – work best with human biology. Dogs can't use shovels. Goats can't put on panty hose. Kittens can't talk like humans. Nor can they use Dutch ovens, play guitars, or drive cars. Human tools either evolve with or are invented by humans. We can duplicate some tools' effects on machines at times and we can train other species to use them to a degree, but full use of human tools requires full humanity – and even then it is not always possible. To use the well-worn metaphor of software, all software is designed for a particular operating platform, such as an OS or a DOS machine. Just so, human language is designed for humans. But that isn't saying much. We need to look carefully at the cogs and wheels of language to understand this design. What ultimately matters is what *we* have to be like to serve as the platform for this linguistic machine. We want to know what underlies our abilities in this sense and how specifically these abilities are dedicated to or designed for language. How do cultures exploit biology to build languages, and how did or does human biology respond to all of this?

There are three platforms found in each individual on which human language rests: the cognitive, the physical, and the cerebral. And when we look at these platforms carefully a curious fact emerges: the platforms are specific to humans but are not specific to language, with the exception of one part of the physical, the shape of our vocal tract.

The physical platform is our body: you need a human body for normal human communication. You cannot speak like a human without a vocal apparatus like a human's, though parrots and some other creatures come close.

As part of our bodies, our brains supply two subtypes of platform, the cerebral and the cognitive. Ultimately these are likely both the same – the brain. But in our current state of knowledge it is useful to separate them in order to discuss their specific contributions. The cerebral is the physiology and organization of the brain that underwrites human language and related abilities. The cognitive platform consists of a set of crucial abilities that distinguish humans from other species.

For example, we need to think in certain ways before we can have language. Language does not enable all thought. In fact, it is partially parasitic on thought. Even though language contributes to human thought, thinking nevertheless goes on in non-human heads also, just as it went on in pre-human hominid heads. In fact, if thought were not possible in some way without language, we could never have achieved language in the first place.

Some scientists say that language might have appeared very suddenly in human evolutionary history, without thought being necessary at all – the result of physics rather than evolution. They reason that once the number of neurons in the human brain passed some unspecified threshold, syntactic processing – ordering and grouping of bits of information – might have just 'happened.' As these folks rightly remind us, we just do not know what kind of chain reaction the compression of $10^{10}$ (100 billion) neurons, with all their attendant axons, dendrites, and synapses, into a brain the size of a grapefruit might have caused. We cannot rule out the possibility that our brains' growing evolutionary complexity somehow gave us the ability to sequence and group information as a by-product. I don't find this a very interesting proposal with which to begin an investigation into the origins of language, though, because we already know that natural selection has the ability over time to add complexity to organisms by rewarding successful novel behaviors, anatomy, and physiology with more offspring, or by punishing non-adaptive behavior with fewer offspring. It is therefore more economical and useful to pursue natural selection as the source for the language platform before moving to alternative explanations, such as physics.

Let us begin discussing the platforms by addressing some of the cognitive abilities that underwrite language. The first crucial cognitive capacity is intentionality. Pirahã men, like many aboriginal people of the world, fish with bows and arrows. Bow-fishing takes considerable skill. I learned this first-hand one day a few weeks into the dry season when some Pirahãs invited me to go fishing with them. That is unusual. Normally they don't invite outsiders to go fishing, because they don't take our fishing abilities seriously. Also, most outsiders are too large for their very small fishing canoes, designed for men weighing less than 160 pounds – although they do invite me along occasionally when they want to fish from my motorboat. But this day was different. They were going fishing away from the river, deeper into the jungle. No canoe was necessary. This sounded pretty unusual, so I decided to go with them. We walked back a mile or so up a jungle stream and then cut inland for a couple of hundred yards, until we arrived at a small body of water that had been left behind by the receding flood water.

One of the dozen or so men who had just arrived pulled out a bundle of timbó, a toxic but usually non-lethal jungle vine, and beat it with a club against a log. He then entered the water with the resultant fine cellulose mesh and pulled it behind him. Brown rivulets of water streamed off from the timbó. He towed his bundle around the entire small pond. Within a few minutes stunned fish started to float to the surface – large dark fish with sharp teeth, called **kaaoxoi** in Pirahã, *traira* in Portuguese, and 'wolf fish' in English. The other Pirahã men circled the little pond and stood with their bows drawn, long, sharp fishing arrows ready to fire. As the fish rose to the surface, everyone started shooting – a silent ichthyic massacre. I told the Pirahãs that it looked difficult to shoot fish with arrows. They replied, 'No. Children can do it.' And one of them gave me a bow and arrow. I proceeded to provide them with a good thirty minutes of hearty laughter before giving up, shaking my head and handing back the bow. They even threw a fish on the ground for me to have a go and I stood practically on top of it in my attempts to shoot it. But for me it was almost impossible to get the arrow to pierce the fish rather than slide off its scales. Pirahã men do it first time every time.

Let's think about what occurred on this fecund fishing trip, so rich in cultural, linguistic, and cognitive information. It teaches us much about cognition. It may initially appear trivial, but something here constitutes a crucial precondition for our ability to communicate. The Pirahãs all and individually directed their thoughts towards several things on this fishing expedition: a goal (fishing); a geographical location (the place in the jungle where the pond was located); and some physical objects (the fish, the pond, and the timbó, among other things). The directedness of their thoughts is shown by the fact that we all went fishing, arrived at the pond, and arrowed fish – well, some of us did anyway. This ability to direct our thoughts is essential to being able to think or talk. And the study of the nature of this directedness has a long pedigree in philosophy.

The cognitive capacity for directedness is called intentionality. Intentionality is how we attribute meaning to the signs of language, by making them *about* or *directed at* something. This directedness is the core of meaning and, as some philosophers would have it, the essence of our mental lives.

Jeremy Bentham, the Enlightenment thinker and godfather of University College London – whose stuffed skeleton and wax head are still on display at UCL as a physical declaration that there is no resurrection – was the first to talk about the concept of intentionality as the hallmark of our mental lives. He claimed that our thoughts, beliefs, desires, and so on, are directed at something. Subsequent philosophers such as Franz Brentano and Edmund Husserl, working in the late nineteenth and early twentieth centuries, further developed ideas on what directedness means for the existence of our mental lives.

Husserl was the first to recognize that *all* of our thoughts are directed at something. Every thought we have has an object, whether that is an object of the real world or a fictional one, and so every one of our thoughts is intentional, according to Husserl.

In more recent times, John Searle, professor of philosophy at the University of California at Berkeley, has worked out a detailed theory of intentionality in human mental life. Searle claims that intentionality cannot be reduced to a brute physical process, but that it is the crucial feature and sign of our mental life. If intentionality is a basic building block of mental existence and if it is not reduceable to the physical, then neither can the mind be reduced to the brain. This is profound stuff.

The ability to have directed thoughts is believed by many philosophers to be essential both for consciousness and for language. Language is, after all, the transmission of directed thoughts from one thinker to another. Consciousness is partially the directing of our thoughts toward ourselves.

When we stop suddenly to listen to a sound in the middle of the night in the basement, our listening is directed, it is intentional. This shows that we clearly can have directed actions and thoughts without language. In fact, at the initial moment of focus there is rarely any language involved. Animals do this routinely. A squirrel that stops eating to look around when it hears a noise is directing its thoughts, thus showing that its thoughts, like ours, are intentional. Searle makes a case in his 1983 book *Intentionality: An Essay in the Philosophy of Mind* that animals (his dog in particular) have beliefs and desires, the prototypical intentional states that underlie much of human behavior. Intentionality is the fundamental cognitive foundation for talking, thinking, and acting. Our talking has a point. Even if you just say 'Good morning,' there is a point to that – you are being friendly to your neighbor – a purpose to which the utterance is directed. Intentionality also comes in at another, more subtle level when we talk. If I utter the word 'dog,' I am directing my word to you, let us say, but the word itself further directs both our thoughts to canines. In this sense, the words and phrases of languages are what I call 'congealed intentionality.'

The notion of intentionality has many specific applications in the understanding of thought and its relationship to language. Just about any sentences can give an idea of what is going on:

*Are you going to the store?*
*You are going to the store.*
*Go to the store!*
*I wish you would go to the store.*

The question, statement, command, and wish each reveal crucial facts about the dependence of language on intentionality. First, they are all directed to 'going to the store.' In the statement 'You are going to the store,' we expect that our thoughts fit the state of affairs we see in the world. That is, we expect that what we say is actually what is happening or what exists in the world. This is called the 'direction of fit' of our words. Whether you are a Pirahã, a German, or an Inuit your language is directed and it must enable a 'mind-to-world direction of fit' – your thoughts must be able to fit the world. A question is an inquiry into the state of the world and thus it shares the same direction of fit as the statement – that is the basis of all information we share in the form of statements; it is a fundamental cognitive prerequisite for language. In the command and the wish, however, we see a different direction in the flow of thoughts and communication. These express our desire to change the fact that the world is not how we want it to be. We desire for the world to fit our wants, instead of remaining the way it is. In the command, I order you to change the world in a certain way by going to the store. In the wish, I tell you that it is my desire that you change the world in a certain way. So our words and thoughts do not fit the world in commands and wishes, they reverse the fit of statements to the world.

The command and wish direction of fit is therefore world-to-mind, rather than mind-to-world. This is simple enough conceptually, but it is none the less vital for communication and thought. Intentionality is thus a fundamental component of the cognitive platform for language.

It is clear that intentionality precedes language since we are not the only species to have intentionality, but we are the only species which has developed language. This ability is essential to survival and so it goes back very far in evolutionary history. Humans also do seem to be the only species to possess some of the other cognitive, physical, and cerebral components of the language platform. But what we share with other species likely originated before *Homo sapiens* existed. The building blocks of language to which I now turn seem to have come to exist after hominids came on the scene.

Although the importance of direction of fit in our intentional life (me → world; world → me) is important for language, it is not enough. There is another prerequisite for a functioning human language – the ability to express different forms of intentionality, different purposes. Linguists refer to the difference in the purpose of utterances as 'moods' and philosophers refer to them as 'illocutionary forces.' Questions, commands, assertions/statements, and desires all exemplify different illocutionary forces.

Philosophers are responsible for another important concept for the understanding of the nature of language and its use, what is called the 'background.' Searle has a famous illustration, which he draws from research in artificial intelligence (AI). A programmer feeds data about common cultural activities into the computer and then asks the computer questions about certain events. The computer's answers at times suggest that it is able to reason. Searle observed a demonstration where the computer is told the following:

*A man went into a restaurant and ordered a hamburger. When the hamburger arrived it was burnt to a crisp, and the man stormed out of the restaurant angrily, without paying for the burger or leaving a tip.*

*The computer scientist then asks the computer, 'Did the man eat the hamburger?' To this the computer answers, 'No.'*

Searle then asked if he could put a question to the computer after this display: 'Did the man put the hamburger in his ear?' The computer answered, 'I don't know.'

Now, that is surprising. If the computer had learned about humans and their cultural activities, shouldn't it know that we don't stick food in our ears, by and large? All humans know this. Somehow the computer was not learning things that all humans know as a matter of course. And Searle claimed that it was incapable of doing so, even in principle. Searle called this information, the background. All humans belong to cultures and share values and knowledge with other members of their cultures. Modern computers are unable to learn culture. Therefore, they can never learn a language, though they can learn lists of grammatical rules and lexical combinations. Without culture, no background; without background, no signs; without signs, no stories and no language.

Over three decades ago, I began linguistic research among the Pirahãs. As I learned to speak Pirahã and began to visit their villages, I noticed that some people would stare open-mouthed when they heard their language coming out of my red-bearded white face. Frequently I would ask them a question but, rather than answer me, they would turn to each other and say things like, 'He spoke! He sounds like Pirahã!' They would say these things right in front of me as if I weren't there. Sometimes they wouldn't even answer me at all: they just stood around talking about my use of their language. Initially, I thought that this was very rude of them. But I eventually realized that some of these people were no more prepared to believe that I spoke their language than they would if I were a Chatty Cathy doll. I realized that some Pirahãs looked upon me as little more than a big, bipedal parrot. You wouldn't answer a parrot if it asked you where you were going, would you? You'd just say, 'Damn, that sounds like a person.' Dan the parrot, that's me.

Part of the explanation for the Pirahãs' behavior towards me stems from the next piece in the cognitive platform for language, the theory of mind. It is worth discussing here because a theory of mind is vital for acquiring language.

The word 'mind' is a way of talking about the cognitive functions of the brain that we do not fully understand physiologically at the present.

The brain's role in thought is not directly observable, though some recent experiments using sophisticated brain scanning equipment have brought us a bit closer.

Since we cannot see other people thinking, we cannot prove that they are not programmed parrots. But we can have a testable theory that they have minds and that they do think, a theory based on their interactions with us, as well as on our other observations about them and ourselves. A theory of mind is vital to communication between us and others. If someone believes that they have a suite of cognitive abilities that underwrites their own intentions, then they have a theory of their own mind, a theory reached through introspection. But if they believe that other creatures have minds more or less like their own, then they can be said to have a theory of mind in a more general sense. Most researchers believe that only humans possess a theory of mind. However, whether other species might have forerunners to this is still in need of research.

Now, the reason that human language requires speakers to have a theory of mind is that communication is predicated on the belief that what is in one mind can be transmitted to another. People believe that they can communicate their thoughts to others because those others of their species are able to understand them – precisely because they have similar thoughts. And to illustrate this, I can think of no better example than humor, especially the finely developed humor of some professional comedians.

The late Richard Pryor, in my opinion one of the best comedians the US ever produced, made me laugh because his humor revealed that he and I had similar minds, that we had similar fears, thoughts, and ways of reacting to the world. He helped many of us to see into other minds. When Pryor offered a brilliant impersonation of an Italian mobster he once worked for, we all laughed because of Pryor's genius, a genius based on his uncanny ability to reveal other minds – his own mind, the mobster's, and his audience's.

The theory of mind helps me to understand why many Pirahãs used to stare at me (some children still do) and talk about me in front of me – they didn't believe I had a mind! At least not one like theirs.

These two parts of the cognitive platform, intentionality and the

theory of mind, are both necessary for language but are not by themselves enough. We need a few more pieces for our cognitive platform
yet. There are three others that stand out: the capacity to distinguish
'figure' from 'ground;' the ability to make contingency judgments; and
consciousness.

The ability to distinguish figure from ground is an ancient feature,
like intentionality. As well as being part of our cognitive platform for
language, it is an ability that all creatures must have to survive. We use
this ability all day every day; our lives depend on it. Here is an example
from the Amazon. A Banawá man, one of the last speakers of his language on the planet, goes hunting in the jungle with his 8-foot-long
blowgun and a woven quiver of poisoned darts. Hearing a sound, he
looks for its source (intentionality). He is surrounded by the vegetation
of the rain forest, his surroundings composed of trees, spaces between
trees, other plants, and animals. In the upper parts of the trees there
are leaves and branches. On several branches of contiguous trees, the
Banawá man picks out against the background of verdant forest several
black human-like shapes – wooly monkeys! What must he do in order
to recognize the monkeys? He must perceive figures (the monkeys)
against a background (the vegetation of the rain forest). Distinguishing figure v. ground is partially dependent on intentionality, but it is
a complex ability which also depends on other functions. In order to
survive in the rain forest the Banawá man needs to be able to focus
on some objects and determine their relationship to the world around
them – moving, stationary, receding, approximating, and so on. Once
we can distinguish figures v. ground we can recognize specific objects
from among other objects and we can focus our attention and intentions – crucial abilities for consciousness and communication.

Contingency is an additional ability that forms part of the cognitive
platform for language, though it is partially dependent on the figure–
ground ability. Contingency is the ability to see links of causality or correlation between objects, such as that if a car is coming quickly and a
child is about to step into the road, he or she might be in danger. And it
is an important source of our ability to theorize. The theory of mind is
dependent on this, for example. We see connections and we construct
beliefs, using our intentionality and ability to distinguish figures from

grounds, and these beliefs form our theories of the world. Many people have studied and written about contingency because of its importance to so many areas of human cognition. Of course, the ability to make contingency judgments, that is, to form the belief that x and y are related in some way, does not mean that our judgments are always good ones.

For all cultures, some contingency judgments can be highly unscientific and wrong, yet nevertheless still play a useful role. For example, in 1981 I was living in the beautiful city of Campinas, in the state of São Paulo, Brazil, studying at the Universidade Estadual de Campinas for my PhD in linguistics. I had just returned to the city having spent nearly a year with my family among the Pirahãs. One day I received a letter postmarked from the tiny Amazonian town of Humaitá. The letter was written in a shaky hand, in Portuguese. I was surprised at first because I did not think that I knew anyone in that city. But then I saw that the letter was from my friend Godofredo Monteiro, a river trader living at a little settlement called the Auxiliadora (the full name is Nossa Senhora Auxiliadora, 'Our Lady the Helper'). Humaitá was the closest post office, a two-day boat trip away from his settlement. In this sad letter, Godo – as his friends called him – told me that his twelve-year-old daughter, Sonia, had died. He said that she and a friend had been struck with violent abdominal pain, vomited horrible things, and died within seventy-two hours in heart-wrenching agony. I felt wretched because Sonia was about the age of my oldest daughter, Shannon, and they had been friends. Godo and his family visited us from time to time at our thatch-roofed hut in the Pirahã village and we always hired Godo's boat to transport us from the village to the Madeira river, where we could catch a large passenger vessel back to Porto Velho, where there was an airport. Godo went on to tell me in his missive that they had discovered what had caused Sonia's death – she had 'misturou as frutas' (mixed her fruits). She had eaten a mango and a pineapple and drunk a glass of milk. Godo concluded that this deadly concoction had killed her. I learned that many of the poor of Brazil had similar beliefs.*

---

* Another was that if you took a bath immediately following a meal, this was likely to kill you or make you very ill (similar to the equally spurious western notion that you should not go swimming right after eating).

Forming incorrect beliefs based on mistaken contingency judgments can lead us to avoid healthy foods or to live unnecessarily sheltered and restricted lives. In some cases bad contingency judgments can kill us, such as by believing that we have better health if we expose ourselves directly to solar radiation or eat lots of saturated fat. Overall, though, even with a relatively high error rate, the human cognitive tool of contingency judgment works in favor of our survivability. Those with the best contingency judgments are more likely to transmit their genes to future generations than those with weaker judgments, thus improving the species.

Our contingency judgments have linguistic applications as well. We learn not only to connect objects to objects, objects to events and events to events, but also words to words. We learn to predict, by means of what some researchers think is a sophisticated and unconscious computational calculation of probabilities, what a speaker is likely to say next, once we learn that the relationships between words are contingent and what the likelihood of one word following another is. For example, in English if you hear a definite or indefinite article (the words 'the' or 'a'), you will expect to hear next an adjective or a noun. Alternatively, if you hear a noun followed by an adverb, you will expect to hear next an adjective or a verb, as in 'I quickly __,' where we expect the blank to be filled by a verb like 'run,' not a noun like 'house.' This is very useful. The statistical generalizations that humans reach are based both on the most common word orders in their language and on what people in their culture usually talk about. Contingency judgments enter a language virtually as soon as it comes into being or someone begins to speak it. As part of our general cognition and cultural experience, thinking in terms of contingency helps us to construct a theory both of what people of our society will say and how the parts of what they say fit together.

Background, culture, intentionality, contingency, and figure v. ground recognition provide us with an extremely rich cognitive platform. But these all depend in some way on the further human capacity of consciousness, to be aware that we are speakers engaged with other speakers, to be aware of our lives and of our surroundings, and that we should focus this awareness on the communicative task at hand. Only conscious beings can communicate. Some researchers argue that consciousness is only possible because we have language. But this seems incorrect. The famous deaf, mute, and blind nineteenth-century political campaigner Helen Keller learned to speak only after the age of seven yet in her books she talks about being conscious long before she learned any language. She is an obvious counterexample to this theory.

But even dogs contradict the idea that language is necessary for consciousness. Dogs are capable of awareness, though perhaps not self-awareness, and so they are conscious, at least by broader definitions. My dog, for example, is aware that I am around; he focuses his attention on me; and he communicates with me, letting me know when he is hungry, ready to go outside, anxious to sit beside me on the couch, and so on. His consciousness is not developed to the same level as mine, so far as I can tell, but it is a consciousness without language.

Each element of our cognitive platform enhances the others, having co-evolved over the millennia into the complex mental faculties we now possess; and each component is necessary for the functioning of the whole. Being able to formulate thoughts and sentences that are appropriate and intentional requires consciousness. If you promise me that you will come to my house tomorrow, you need to be aware of me, my house, what you are saying, and how I am interpreting what you are saying. Whatever consciousness is and whatever creatures can be said to possess it, it is a crucial component of the cognitive platform for language.

The final component of the cognitive platform I will discuss here is in a way a by-product of these other features: culture. Culture is essential to language, in the sense that it results from networked knowledge and behavior with others and brings *meaning* from the world. This is a crucial component because these features of culture are what motivate and enable us, but are also shaped by the growth of language among groups of our species. So, culture is both a product and a producer of language. Likewise, language is a product and a producer of culture.

Culture is not merely part of one individual's cognition. It is the external cognitive link between multiple individuals. Culture is a set of values that originate in cognitive links produced by the interactive relationship between members of a society. The values and their rankings each reside *in part* in the cognition of each member of that society. But only the society as a whole possesses all the values and knowledge that lie within it. Each individual possesses a larger or smaller subset of the entire society's values and knowledge depending on that individual's experiences and intelligence. These internalized values and knowledge then serve as tools to help the individual know what to do, what to expect, how to react, and so on, to the varied dangers, pleasures, and other experiences of life without having to develop an entirely original response to each stimulus. These tools enable us to spend less time worrying about how we should live and what we should do and spend more time living and doing.

I used to take long walks with the Pirahãs in the jungle. As I traveled along, I would take out a pen and notebook to write down the words for objects they would tell me about during our journey. Then I realized an embarrassing fact – I did not know the names in my own language for the flora and fauna that the Pirahãs were trying to teach me about. My culture contains all of this information, but most of it is stored outside of my brain, in textbooks, in the brains of specialists, on the internet. Cultures are useful for many reasons. But it was only when I lost access to mine for a while that I realized acutely that I was but a tiny point in a cultural network, unable to fully function apart from my culture.

Since biologically we humans can only access information through our five senses, we must have a way of transmitting language to each other's senses or language would not be possible. This is part of the basic communication problem as Shannon outlined it and any communication system needs to solve the transmission and reception problems. As biological creatures we must be able to use our bodies and brains and transmitters and receptors to produce an understandable message. Therefore, we need to make sure that the people we want to talk to use the same symbols as we do, and use them in more or less the same way (this is discussed in detail in the next chapter). You and I both need to agree that 'white' means white and 'up' means up, for example. The second issue is that we need a way to get a symbol into a medium (water, air, line of sight, etc.) that will get the symbol to the senses of our addressee.

With regard to how humans are able to transmit and receive their own symbols, we have two main choices, each with a transmitter and a receiver: hands and eyes or mouths and ears. There are other ways, such as colored flags, smoke signals, Morse code, typed letters, chicken entrails, and other visual means. But, funnily enough, no community has ever been found that communicates internally exclusively by writing or smoke signals unless its members have some sort of shared physical challenge or are all cooperating with some who do.

Hands and eyes are often used, however. Among the Urubu-Kaapor people of north-eastern Brazil, for instance, most members of several Kaapor communities know sign language. In 1968 my friend the linguist James Kakamasu observed that seven individuals out of the total Urubu population of about 500 were deaf. But all members of the population were fluent in their own sign language and all used this sign language as their principal channel of communication whenever any of the seven deaf members of the community were present. This shows the use of sign language as almost a primary medium of communication for a group and illustrates, once again, that language is utilitarian.

As I showed in chapter six, though, humans prefer to use their voices rather than their hands as the primary channel of communication. So what is it about human anatomy and physiology that makes the use of the human voice a viable transmitter of language? It is time to look more carefully at the physical platform for the human voice.

Whereas other primates have a small, fixed number of calls, humans have veritably unlimited, non-finite languages. Human speech, combining voice, symbols, syntax, culture, and so on, is capable of transmitting an astoundingly high rate of information relative to other animals and modes of communication. Humans evolved to talk, not to write, not to type, not even to speak sign languages. Our anatomy changed for vocal speech.

It was not only our anatomy that changed to facilitate vocal speech, though, but our perceptual abilities. The sounds made by children, women, and men are different. The frequency of an 8-year-old's voice is considerably higher than that of a 38-year-old. And yet, an adult has no difficulty distinguishing a child's words or understanding all of them. The ability to generalize and use different manifestations of what we perceive as a unified domain is not unique to language. We have no difficulty, for example, in keeping our balance while riding *either* a bike *or*

Figure 6 **Face diagram of human vocal apparatus**

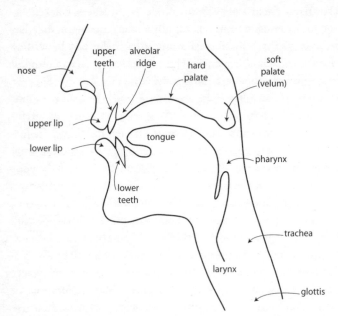

a motorcycle – two related, though radically different, forms of transportation. Humans are relatively good generalizers.

What underlies our wonderful human voices is a jury-rigged collection of anatomical parts that we need for other things. In the 1980s *MacGyver* was a popular American television program about a resourceful secret agent who regularly extricated himself from dangerous and apparently hopeless situations by taking mundane objects from his surroundings and using them for purposes for which they were not originally designed – using an aerosol can to make a bomb, for example. The human vocal apparatus is like one of MacGyver's devices, exploited not merely by some fictional adventurer but by evolution itself. Not one part of the vocal apparatus evolved primarily for language. Each part has a separate function. But out of these disparate parts has evolved a vocal apparatus for speech.

To speak we use our lips, teeth, tongues, nasal passages, vocal cords, and other body parts. The sound *b*, for example, is made by simultaneously blocking the flow of air from the lungs through the mouth by closing the lips and vibrating the vocal cords, parallel muscles that flank the opening to the trachea, the glottis. The sound *p* is identical to *b* except that with *p* the vocal cords (or folds) do not vibrate. The sound *m* is the same as *b* except that the air that originates in the lungs also passes through the nose (which is why when you have a cold and your nose is blocked up, *m*s sound like *b*s). In Figure 6 you can see the physical mechanisms that underlie all of our sounds.

Again, every part of the vocal apparatus has a non-speech-related function that is more basic from an evolutionary perspective than speech and that is found in other species of primates, such as the larynx. In order to produce language we exploit what our bodies and brains already have. Therefore it is not surprising that the neural mechanisms implicated in human language, as well as tongues, teeth, and the rest, such as the basal ganglia, are not only part of the endowment of modern human biology, but are found in many other animals as well. This is a simple consequence of the continuity of evolution by natural selection.

But, as seen earlier, there is one unique aspect of the human vocal apparatus that does seem to have evolved specifically for human speech: its *shape*. Let's first look at how the vocal apparatus does its job, then we can discuss its novel shape.

One question worth asking is whether there is anything special about human speech or whether it is just composed of easy-to-make noises?* Also, would other noises work just as well? Not really. For example, a possible alternative to human speech sounds is Morse code. The fastest speed a Morse code operator can achieve is about fifty words per minute. That is about 250 letters per minute. Morse code operators working this quickly, however, need to rest frequently and can barely remember what they have transcribed. But a hungover college student can easily follow a lecture given at the rate of 150 words per minute! We can produce speech sounds at roughly twenty-five per second.

How does speech perform this magical work? It has structures

---

* For a fuller account of the evolution and essential properties of hominid speech, I refer the reader to Philip Lieberman's *Toward an Evolutionary Biology of Language* (2006), from which I have taken much of the following material.

– subtools – that make language more efficient. One of the subtools human biology and culture have given us for speech is the syllable. Syllables are used to organize phonemes into groups that follow a few highly specific patterns across the world's languages. The most common patterns are things like Consonant (C) + Vowel (V), C+C+V, C+V+C, C+C+C+V+C+C+C, and so on (with three consonants on either side of the vowel pushing the upper limits of the largest syllables observed in the world's languages). English provides an example of complex syllable structure, seen in words like *strength, s-t-r-e-n-g-th,* which illustrates the pattern C+C+C+V+C+C+C (with 'th' representing a single sound). But what I find interesting is that in the majority of languages C+V is either the only syllable or by far the most common. With the organizational and mnemonic help of syllables and our neural evolution plus our contingency judgments based on significant exposure to our native language, we are able to parse our speech sounds and words far faster than other sounds.

Suppose you want to say, 'Piss off, mate!' How do you get those sounds out of your mouth on their way to someone else's ear? There are three syllables, five consonants, and three vowels in the words 'Piss off, mate,' based on the actual spoken words rather than the written words using the English alphabet. The sounds are, technically, [pʰ], [I], [s], [ɔ], [f], [m], [eⁱ], and [t]. The syllables are [pʰIs], [ɔf] and [meⁱt] and so, unusually in English, each word of this insult is also a syllable.

Sign languages also have much to teach us about our neural cognitive–cerebral platform, however. Native users of sign languages can communicate as quickly and effectively as speakers using the vocal apparatus. So our brain development cannot be so tightly connected to speech sounds that all other modalities or channels of speech are unavailable. It seems unlikely that every human being comes equipped by evolution with two separate neuronal networks, one for sign languages and another for spoken languages. It is more probable that our brains are equipped to process signals of different modalities and that our hands and mouths provide the easiest ones. Sign languages, by the way, also show evidence for syllable-like groupings of gestures, which indicate a human predisposition to such groupings, in the sense that our minds quickly latch on to syllabic groupings as ways of better processing parts of signs. Regardless of other modalities, though, the fact remains that vocal speech is the channel exclusively used by the vast majority of people. And this is interesting, because in this fact we do see evidence that human physiology has evolved for speech.

Human infants begin life much as other primates vocally. The anatomy of a baby's vocal tract above the larynx (the supralaryngeal vocal tract or SVT) is very much like that of the corresponding tract in a chimp. When a human newborn breathes, its larynx rises to lock into the passage leading to the nose (the nasopharyngeal passage). This seals off the trachea from the flow of mother's milk or other things in the newborn's mouth. Thus our babies can eat and breathe without choking, just like chimps.

Adults lose this advantage. As we mature, our vocal tract elongates. Our mouths get shorter and our pharynx (the section of the throat immediately behind the mouth and above the larynx, trachea, and esophagus) gets longer. Consequently, the adult larynx does not go up as high relative to the mouth as a baby's and thus is left exposed to food or drink falling on it. If this kind of stuff enters our trachea, we can choke to death. We must therefore coordinate carefully the tongue, the larynx, a small flap called the epiglottis, and the esophageal sphincter (the round muscle in our food pipe) to avoid choking while eating. One thing we must be careful not to do, therefore, is to talk with our mouths full. Talking and eating at the same time can kill or cause severe discomfort. So, as we grow into adulthood we seem to lose an advantage possessed by chimps and newborn humans.

But the news is not all bad. Although the full inventory of changes to the human vocal apparatus is too large and technical to discuss here, the final result of these developments is good for talking because we can make a larger array of speech sounds, especially vowels, like the supervowels 'i,' 'a,' and 'u,' which are found in all languages of the world. These are the easiest vowels to perceive. We are the only species that can make them well.

This evolutionary development of the vocal apparatus gives us more options in the production of speech sounds, a production which begins with the lungs. The human lungs are to the vocal apparatus as a bottle of helium is to a carnival balloon. Our mouths are like the airpiece on the balloon. As you let air out of a balloon, you are able to manipulate the pitch of the escaping air sound by relaxing the air piece, widening or narrowing the hole through which the air hisses out, cutting the air off intermittently, and even 'jiggling' the balloon as the air is expelled.

But if our mouths and noses are like the balloon's airpiece, we have more moving parts and more twists and chambers for the air to pass through than a balloon. So we can make many more sounds than a balloon. Nice to know. And since our ears and their inner workings have co-evolved with our sound-making system, it is not surprising that we have evolved to make and be sensitive to a relatively narrow set of sounds that are used in speech. Hearing and producing speech sounds are two vital evolutionary adaptations found universally in all healthy humans.

There is not much more to say about the lungs' relationship to speech here, since this is not a phonetics textbook. Lots of creatures have lungs. But as we go up the vocal apparatus from the lungs, we run into our first hominid innovation in the vocal apparatus, the larynx.

Some researchers have argued that only *Homo sapiens* possesses the anatomy to support speech as we know it. They claim that not even *Homo neanderthalensis*, our closest relative, had this capability. They might very well be correct. But whether *Homo neanderthalensis* had a simpler vocal apparatus or not, it would be a mistake to confuse the evolution of the ability to have *spoken* language with the possession of language.

According to evolutionary research, the larynges of all land animals evolved from the same source – the lung valve of ancient fish, in particular as seen in the Protopterus, the Neoceratodus, and the Lepidosiren. Fish gave us speech as we know it. The two slits in this archaic fish valve functioned to prevent water entering into the lungs of the fish. To this simple muscular mechanism, evolution added cartilage, and tinkered a bit more to allow for mammalian breathing and the process of phonation. Our resultant vocal cords are therefore actually a complex set of muscles. They were first called *cordes* by the eighteenth-century French researcher Antoine Ferrein, who conceived of the vocal apparatus as a musical instrument.

The hard part of the front of the throat is the trachea, or 'windpipe,' which connects the larynx to the lungs. A series of interesting little bones holds the larynx suspended to guarantee that it works properly for phonation. Phonation is the activity of the larynx in modifying the airflow by stretching, relaxing, and vibrating. These functions are produced both by muscle movement and by the alteration of pressure above and below the vocal cords.

Although the lungs are the only air source for European languages, the airflow of human speech does not always originate there. For the languages descended from the ancient Mayans – those spoken by peoples like the Tzotil, Tzeltal, Ch'ol, and others – so-called 'glottalized' sounds (implosives and ejectives) are common. When I began my linguistic career, in the mid-1970s, I went to live for several months among the Tzeltales of Chiapas, Mexico. One of my favorite phrases of theirs was **c'uxc'ajc'al** 'It's hot outside,' which contains three glottalized consonants (indicated in Tzeltal orthography by the apostrophe). To make these sounds, the glottis – the space between the two vocal cords in the larynx – must be closed, cutting off air from the lungs. If the entire larynx is then forced up at the same time that the lungs or tongue cut off the flow of air out of the mouth, then pressure is created. When the tongue or lips release air out of the mouth, an explosive-like sound is produced. This type of sound, seen in the Tzeltal phrase above, is called an 'ejective.' The opposite of an ejective is an 'implosive' sound. To make an implosive, the larynx moves down instead of up, but everything else remains the same as for an ejective. It is this downward motion of the larynx which produces an implosive – caused by air suddenly rushing into the mouth. We do not have anything like these sounds in English and I remember having to practice ejectives and implosives constantly for several days, in order to pronounce them correctly, since the Tzeltales I worked with use them both. They are interesting sounds – not only are they fun to hear and make, but they extend the range of human speech sounds beyond the strictly lung-produced sounds in European languages.

A different type of glottalized sound worth mentioning is produced by nearly, but not quite, closing the glottis and allowing lung air to barely flow out. This effect is what linguists call 'creaky voice.' People often produce creaky voice involuntarily in the mornings after first arising, especially if their vocal cords are strained through yelling, drinking, or smoking. But in some languages, creaky voice sounds function as regular vowels.

Air pressure for speech sounds can also be produced by the mouth, with a little help from the larynx or the lungs. Sounds produced in this way are called 'clicks' and they are common among languages of Africa, such as Xhosa, the language of Nelson Mandela, and other Khoisan languages. We actually use a limited range of clicks in English, as in the 'tsk tsk' sound, or the sounds we sometime make to get a horse to move, 'klk, klk.' The difference is that clicks in Bantu languages are fully functioning consonants in the language whereas in English they are 'paralinguistic,' just special sounds we use for special functions, but which do not actually appear in English words.

Like glottalized consonants, clicks involve two points of closure. But since the source of air pressure is the mouth alone, the airflow is blocked in the front and back of the mouth. A suction pressure is produced by the tongue pulling air to the back of the mouth. When the closure farther towards the front of the mouth is released, at the lips or the front of the tongue, a click is produced.

Although humans can produce a rich array of sounds, for speech they do not need to. By the use of only a small range of consonants, intermixed with three or more vowels, all human meaning can be communicated. Nevertheless there have been other modifications in our bodies that have been needed to get us to full human speech capacity.

Although there are serious problems with the idea that there are larger, language-specific areas of the brain such as Wernicke's area or Broca's area, several researchers have shown that there are cerebral underpinnings of language in the subcortical region, and these are known as the basal ganglia. The basal ganglia are a group of brain tissues that appear to function as a unit and are associated with a variety of general functions such as voluntary motor control, procedural learning (routines or habits), eye movements, and emotional function. The basal ganglia are strongly connected to the cortex and thalamus, along with other brain areas.

The general nature of the basal ganglia, their role in speech, and their responsibility for habit formation teach us several things. First, these fundamental components of language function are not specifically designed for language, although harm to the basal ganglia can produce a number of aphasic conditions, or language deficits. This means that the responsibility for language lies with various regions of the brain that contribute in multiple ways at a higher level of organization in our mental or cortical life. Second, language is at least partially a series of acquired habits and routines, along with others such as the ability to ski, ride a bicycle, type, and so on.

These habits of our intentionality work together with our societies to help us have language. But how does a community of people sharing a culture make a language? Plato's star student, Aristotle, pondered this very question.

# PART THREE:
## Applications

## Chapter Eight

# ARISTOTLE'S ANSWER: INTERACTION AND THE CONSTRUCTION OF CULTURAL SIGNS

*'A social instinct is implanted in all men by nature ...'*
Aristotle, *Politics* (*c.* 340 BC)

I n their 2009 book *The Interactional Instinct: The Evolution and Acquisition of Language*, a group of researchers from UCLA led by Namhee Lee claim that 'Crucial for language acquisition is what we call the "interactional instinct." This instinct is an innate drive among human infants to interact with conspecific caregivers.' These researchers have rediscovered Aristotle's 'social instinct.'

The authors allow that we may not be the only species to have this instinct but argue that it is more powerful in humans than in other creatures. Several researchers have claimed that the acute dependence of humans on society and their concomitant interactional instinct may result from the fact that human infants are dependent for others on help for their survival and development for an extraordinarily long time.

Primates are at the apex of the mammalian class in several ways. For one thing, they show the strongest bond between mothers and their offspring. Primate mothers, especially the great apes (which includes humans), could almost be said to form a single complex organism with their offspring. Researcher Sarah Hrdy reports that mother orangutans are in constant contact with their young, not losing physical contact for even an instant, for at least the first five or six months of life.

The interaction of mothers and children requires a good deal of mind-reading – each needs to be able to predict what the other is going to do. Children need to know whether their mothers are going to feed them or scold them. Mothers need to know if their infants will be startled or pleased by coming experiences. Mothers and babies want to know each other's emotional and physical states. The ability of each member of a community to predict what the other will do is vital. When the young are able to learn from their mothers and others in their community, their survivability becomes stronger.

When I and my family were living in the Amazon jungle and river traders or others, having been drinking all day, ran to our house, agitated and shouting angrily at us, my children would look at my face. If they saw fear in me, they felt fear themselves. If they sensed that I was calm, not panicked, then neither were they. Babies and children learn from their parents' faces what is in their parents' minds and they adjust their own inner mental lives accordingly. If we're OK, they're OK.

Unlike other primates, however, human mothers allow others in their community to interact with, hold, and feed their babies. Human primates will happily raise one another's offspring in most cases, thus providing all with a greater likelihood of survival past infancy. The young of other primates are less likely to survive the death of their mother. This is due to the rigid relationships they build with their mothers to the exclusion of others.

Hrdy says that 'No mammal in the world has produced young that take longer to mature or depend on so many others for so long as did humans in the Pleistocene ... these incredibly costly, large-brained offspring grew up slowly.' During this prolonged period of development, human infants need help. They cannot survive without interaction.

With social and cultural learning so obviously dominating a child's early years and enveloping its language-learning process, it is surprising to see a growing body of literature that proposes that learning and general intelligence are not sufficient to account for human cognitive and cultural accomplishments. I refer to this body of writing as the 'strong instinct movement.' A plethora of instincts has been proposed in recent years: Steve Pinker's 'language instinct,' Denis Dutton's 'art instinct,' Michael McCullough's 'forgiveness instinct,' Marc Hauser's 'moral instinct,' and Nicholas Wade's 'faith instinct,' among others. But instincts do not work like these authors seem to think. Instincts have zero learning curves. A baby sea turtle does not learn to walk towards the beach; it just does it. A human baby does not learn to suckle or grasp; it just suckles and grasps. A duckling does not learn to imprint on its mother; it just does. Instincts do not have different levels of attainment for healthy individuals. It is not the case that healthy sea turtle babies will be divided into those who crawl part way to the beach and those who crawl all the way. Normal human babies do not divide into those who grasp and those who occasionally grasp. Learning is not necessary with true instincts.

On the other hand, all normal human children do learn to talk. Perhaps it is the case that, although learning to talk is not an instinct, the desire to learn is. In fact, I think that this is how an instinct might indeed play a legitimate role in the acquisition of human language. But this instinct itself is neither learned nor nearly as all-inclusive as the language instinct that others have proposed. Let's not forget that language, art, faith, and morality all have learning curves and different levels of attainment. And none of them are simple actions or developments that are of the type commonly associated with instincts. They are complex behaviors, learned in every sense of the word, and moderated in the developing child by culture. There is no argument that any of these come about unlearned.

The interactional instinct, on the other hand, requires no learning. It is a simple reflex of the type more commonly associated with instincts. It is much more plausible as a reason for the development of language than, say, a full-blown language instinct. The role of this instinct comes out of the need to communicate which forces all humans to learn a language, and to invest the effort into learning a language. Aristotle recognized this before anyone else. In fact, so many of Aristotle's philosophical musings on language are relevant today that it is worth thinking in more depth about what he had to say over two thousand years ago.

We know very little of certainty about Aristotle the man. He was born in 384 BC in Stagira, Greece and he died in 322 BC in Athens. He was a student of Plato and a teacher of Alexander the Great. His catholic interests led him to write on nearly every domain of learning recognized in the ancient world. He spent some twenty years studying at Plato's school, the Academy, named after Academus, a hero of Greek mythology. In the wooded area where Plato taught his students, legend has it that Plato recognized Aristotle, forty years his junior, as his most brilliant pupil. Upon Plato's death in *c.* 348 BC, Aristotle said of his master that it was he alone who 'first of mortals clearly revealed, by his own life and by the methods of his words, how to be happy is to be good.'

Aristotle founded his own school, the Lyceum, following Plato's death and took on students in order to continue the tradition of dialogic and syllogistic learning and reasoning of his master. Aristotle's subsequent influence on western culture and civilization cannot be overstated. So influential were his teachings that they were in part responsible, paradoxically, for the Dark Ages, because so many believed that Aristotle had already mastered all knowledge – why should anyone else try to advance it?

The Lyceum was also known as the 'Peripatetic', or 'walkabout', school because Aristotle liked to engage his students while walking and talking. Most of what we have as his writings are in fact notes and memories of his lectures at the Lyceum made by his successors and students.

In his brief reflections on language, 'On Interpretation', Aristotle actually managed to get much of what he said wrong. But no one can ignore it because it was brilliant, pioneering, and systematic. For instance, Aristotle claimed that nouns are words without reference to time, while verbs are identified because they all have a temporal dimension. This is generally true. When we say, 'The Pope is dead', the verb 'is' carries the present tense and the noun 'Pope' has no time meaning in it at all. Yet in many languages there are verbs without time and nouns that do express time. Even English has a few examples that violate Aristotle's claim. When we say something like 'Johnny wants to run,' 'run' is a verb but it has no associated temporal meaning. This is why we call 'to run' the infinitive form of the verb. Some English nouns also present a problem for Aristotle's claim. In a phrase like 'My ex-wife lives far away,' the noun 'ex-wife' refers to a past wife. It is a noun with a past tense.

But Aristotle is important for us in spite of his limitations. The fourth and fifth centuries BC were a fecund period of world history. Aristotle was a towering figure in a world-class group of thinkers in Greece, India, China, and Persia. But we respect Aristotle because so many of his ideas have reverberated down through history – their importance has stood the test of time. Aristotle said, for example, that

'Every sentence has meaning, not as being the natural means by which a physical faculty is realized, but, as we have said, by convention.' By making this statement, he also went against the current of Greek thought, including the teachings of his mentor, Plato. Plato had taught, according to some interpretations of his work, that meaning resided in the true and natural forms of language that existed in heaven. Aristotle stated instead that meanings are conventional. Conventions emerge from values that are sustained and developed by cultures, so Aristotle would agree that meaning is cultural.

Aristotle believed that the major aspects of word meanings and sentence meanings are set by society, part of humans' collective effort to satisfy our 'social instinct.' There can be no culture without language, no language without culture, and no society without both. Language and culture are not a chicken–egg problem; their relationship is more like fire and heat.

The learning of language and culture seems to begin in the womb. This may be why infants appear to be able to recognize their mothers' voices at birth. We never stop learning our language, art, forgiveness, morality, or faith, and we all achieve different levels in these areas – all of which ought to produce strong skepticism towards the claims that these are instincts. The interactional instinct leads to the learning of language. It is what kick-starts human communication.

But the interactional instinct is not the only language-relevant element that is present in our brains at birth. When we are born, we do know some things. We are not a blank slate. However, this knowledge is not a priori, platonic knowledge. It is learned – in the womb. Our gestational information intake includes things like the basic intonational patterns of our mother's voice, how our mother reacted when we kicked, how her body reacted to loud and sudden sounds, a lot about her diet, and other physiological processes. This learning all precedes the dramatically accelerated learning that begins after we emerge from the birth canal.

For most of us, our primary, perhaps only, caregiver was our mother. Our mother's role is strong and vital throughout life. Even at the end. My uncle Aundon fought in World War II with the Marine First Division, the 'Old Breed,' at Peleliu and Okinawa. He won the Purple Heart twice, once for being shot in the knee and having his lower lip shot off and again for being nearly shattered by exploding ordnance from Japanese artillery. He was one of the first on the beaches, in the Marine first wave. He operated a flame-thrower, or 'jelly-gas gun', attacking Japanese machine-gun nests. Today he is a quiet man who does not like to speak about war. But when he does choose to talk, what he says can rend your heart. He saw many men die, and my uncle tells me that the most common last word he heard from these men – none much older than twenty – was 'Mama.'

There is a significant degree of agreement among theoreticians about some aspects of the learning of words. Words are not instincts. They must be learned. Researchers have claimed that the rate at which children learn words is remarkable. It is true that children go from a vocabulary of about fifty words in their first eighteen months to about 350 in their next six months, by the time they are two, and by thirty months they are up to about 600 words. Their rate of learning continues to increase so fast that it must come from some innate ability that speeds up and enhances learning in the early years. On the one hand, this has to be right in some general way. Children do learn words very well – they learn their meanings, their sound structure, their morphology, and their syntactic quirks – and they seem to do all of this quickly. If there is something very unusual about the speed with which children learn words and language, then might this result from their interactional instinct working with their larger primate brains? However they do it, it seems that they also use the same strategy to learn the sentences of which words are a part.

For the last twenty years or so a new theory of grammar has been attracting more and more adherents from the worlds of psychology and linguistics. This idea is that a major part of the syntax of human languages is not built on rules, but instead on core constructions that lead to a family of similar constructions. This new theory is called Construction Grammar and, according to construction grammar, words and most sentences are learned in a similar manner, since both are signs.

Babies learn many aspects of their mother's culture. They begin their progress towards mastery of language as soon as they are born. Nativists believe that mothers trigger and shape the baby's genetically provided grammar and that most of the rest of the linguistic development comes from the words that the infant learns and our genome.

But is children's internal learning machine so specialized or does it serve a more general purpose? Is it an all-purpose device or specifically designed for language acquisition? That is the crucial question in understanding the role of culture in language development. Children learn many things well. Are they filled with multiple little learning programs or do they have only a few, or even just one? Do these learning programs shut down around puberty or do adults continue to learn pretty much in the same way as they did as children?

The debate surrounding these questions is intense and the vast body of literature on the subject of learning is daunting. But most researchers agree that words do offer clues as to how specific our syntax is. And words have to be learned one at a time, however it is that we do this. We cannot, for example, guess what the word for 'dog' is going to be in our language just because we know the word for 'cat.' Word-learning entails rote memorization, the item-by-item mastery of individual signs.

At the same time that children are learning their first language, they are learning lots of other things, from cultural knowledge and values to contingency relationships ('if this, then that'), who their relatives are, how animals behave, how to sit in a canoe, and so on, depending on their particular culture community, their parents' interaction with them, and so on.

Kids are tough, smart, and resilient. They bring this to their biggest job – growing up. I am reminded of what the instructor said at prenatal classes I attended to prepare for the birth of my second child: 'Love your children. But they are tough little creatures. You could keep them in a dark room all day and feed them raw meat, but so long as they know you love them, they'll probably turn out OK.' That is hyperbole. And it was intended to be humorous. But it does express one thing well: kids are good at life. General mental fitness explains much of their ability to learn most of the skills they need, including language.

What word-learning by children *really* has to teach us is that the way they learn signs might be all they need to learn their entire language. Some who believe that language is innate base their arguments partially on the idea that sentences are acquired differently from the way that words are learned. This is because the number of words is finite, but the number of sentences is potentially infinite. However, construction grammar looks at sentences in a different way and says that they are signs, just like words. An entire sentence, like a word, is learned as a single sign – an accepted structure or 'construction' in a particular culture – rather than the output of a formal grammar. That is, we learn many if not most sentences as part of a family of related structures, such as 'The more I eat, the fatter I get,' 'The hurrieder I go, the behinder I get,' 'The bigger they are, the harder they fall,' rather than as the outputs of formal rules, such as S → NP, VP.

This theory takes seriously the idea that the sign is the key to language use and learning. Sentences all share a common set of structures in each language. Think of the family of structures related to an English sentence like 'The bigger they are, the harder they fall.' From this we can derive many other sentences, by analogy, such as 'The higher they go, the farther they plummet,' 'The dumber you are, the higher you can climb,' or 'The hurrieder I go, the behinder I get.' The last example is scientifically interesting. We accept this sentence and are able to interpret it without difficulty because we see it as related to 'The bigger they are, the harder they fall.' And we accept it even though 'hurrieder' and 'behinder' are not accepted by some as grammatical words in English. But they are understood as words because of where they appear in a familiar construction.

Several recent books show that many, perhaps all, sentences are learned like words, with the help of analogy in building families of constructions.* This leads to an interesting debate. Some nativists use sentences like 'The dax gorped the wug' to argue that language acquisition must use abstract linguistic rules. This sentence has nonsense words in the position of the object, subject, and verb. There is no such thing as a 'dax', no action 'to gorp,' and no object called a 'wug.' But children tend not to have any problems decoding such a sentence. They interpret it to contain two nouns and verb that they have not yet learned, and assume

---

* See Michael Tomasello's *Constructing a Language* (2003) (the 'constructing' in his title is an allusion to construction grammar) and Peter Culicover's *Syntactic Nuts* (1999) for two technical studies of how sentences and words are learned in similar ways.

that it must mean something like 'a thing called a dax did something called gorping to a wug.' Nativists use this type of sentence to reason that grammar learning involves learning abstractions along the line of the rules we examined earlier, of the type S—> NP VP; NP —> 'the wug,' 'the dax;' VP —> V NP; V—> 'gorp.' Learning such rules straight-forwardly underwrites the acquisition of new forms.

But there are alternative explanations based on generalizing directly from observed constructions, rather than abstract rules. Rather than learning algebraic procedures for combining symbols, children instead seem to learn linguistic categories and constructions as patterns of meaningful symbols. A common pattern in English, for example, comes from constructions of the form *X VERBed Y*. This construction signifies an action in which one thing affects another in a way speci-fied by the verb. Children learn these signs early on in their exposure to English. The 'X' and 'Y' signify things of some sort. We know that children have to learn words one by one but that they also learn some sentences in this way, as idiomatic expressions. Studies show that in fact there is no other learning needed for most sentences. So a sentence with novel words, like 'The dax gorped the wug,' provides no evidence at all for either an abstract grammar or nativist principles.

Again, the hypothesis that children can learn sentences as signs or symbols, just as they learn words, raises an important question. If so many sentences and all of our words are learned one at a time, just as we seem to be learning everything else we know, is there really any need for an extra tool, an innate language instinct or universal grammar? A language instinct of the kind popularly described in the nativist litera-ture would be a costly investment for evolution to make if we already had the ability to learn languages via our pre-existent intelligence and other abilities, such as sequencing of actions. In other words, if we already have the ability to learn languages without universal grammar, why would evolution give us universal grammar? It would not. Thus before anyone can make a case for universal grammar, they must first present a case against any other form of language learning.

Some leading researchers believe that indeed nativist ideas like a lan-guage instinct or universal grammar have passed their sell-by dates. Michael Tomasello, the Director of Psycholinguistics at the Max Planck

Institute for Evolutionary Anthropology in Leipzig, says exactly this. A world leader in the study of cognitive development in canines and primates, including humans, he says simply 'Universal grammar is dead.' It was a good idea. It didn't pan out.

And part of the cause for its demise has been the recognition that children learn sentences like they learn other signs. This interests me especially because constructions are cultural objects. Just as 'red, white, and blue,' 'kick the bucket,' 'I'll be gol-darned,' and others reflect a specific cultural heritage, a set of cultural values and cultural knowledge, so are the vestiges of cultural knowledge, the *background*, found throughout our grammars.

The work that a universal grammar would have left for it to do after we learn our signs as we learn everything else would be fairly minimal. Thus it is unlikely that, especially given the effort and cost to natural selection of endowing us with this apparently little-needed ability, we would have evolved such a grammar.

The idea that languages are cultural artifacts has been around for a long time. And it enjoys a great deal of support from linguistic and cultural researchers around the world. One entire theory of grammar, known as Cognitive Grammar, recognizes, according to its founder, Ronald Langacker, 'cultural knowledge as the foundation not just of the lexicon, but of central facts of grammar as well.'

To understand what cognitive grammar is getting at here, we first need to understand what it means to have cultural knowledge. Living in a culture and acquiring cultural knowledge enables us to gain meaning from the world around us and from each other. Little things can take on a range of meanings, depending on our culture. So, touching the forefinger to the thumb, while bowing the forefinger to produce a circular space between the thumb and finger means 'A-OK' to most Americans. But it means absolutely nothing to the Pirahãs. In still other cultures, such as in some parts of Brazil, it can be a vulgar sexual reference. There are many ways of living in this world. Diversity of cultures, like diversity of genes, is fundamental to our survival as a species. Hence there is plenty of it.

Let's look at another example of culture shaping symbols. The phrase 'kick the bucket' means to die in English. It may have originated from an expression uttered at hangings – to kick the bucket out from under a condemned person with a noose around their neck. Whatever its origin, though, it clearly originates and is used by speakers that share a common cultural-linguistic heritage – English.

Even everyday expressions drip culture. Take the straightforward 'He is a traffic cop.' The cultural role of a traffic cop is responsible for the existence of the phrase 'traffic cop.' Without traffic, a cultural construction, and without cops, another cultural construction, there could be no word or phrase for traffic cop. Such examples further illustrate how cultural knowledge and values can lead to new words or meanings.

Once we have meanings and concepts, most of these are turned into signs, principally words. Words are stored in every normal human's long-term memory. And so are the meaning relationships between them. There is evidence, for example, that word meanings are stored in an organized way, cross-linked to other word meanings. For instance, when we hear a story about a single farm animal, such as a sheep, psychologists have discovered that several other words for farm animals – horse, cow, pig – are simultaneously activated in our minds, depending on our cultural conception of farms and the animals normally found on them. We can refer to this part of our long-term memory as our mental dictionary or lexicon. And no one should be surprised to learn that our lexicon is partially structured by our culture, because words come from cultures.

But what about the way we structure our grammars – the principles that govern how sentences and phrases are formed, rather than merely the words from which they are composed? Can grammatical structure itself be shaped by culture?

Among the growing number of studies on culture's hold on grammar is one done by John R. Roberts on Amele, a language spoken in New Guinea. Amele is interesting because of the way that it distinguishes and expresses the concept of giving, compared to other events in the culture. Normally in Amele, the subject comes first in the sentence, followed by the direct object and then finally by the verb and its suffixes. In Amele, speakers would utter the equivalent to 'John an apple ate', rather than 'John ate an apple', as would speakers of Japanese, Pirahã, and thousands of other languages. What is surprising and instructive about Amele, though, is that when the meaning of giving is expressed, there is no verb. Here is a typical sentence in Amele:

**Jo    eu       ihaciadigen**
*house that    show*
'I will show that house to you.'

In this example, there is a verb, **ihac**, 'to show'. But in the next example, which is about giving, there is no verb:

**Naus Dege houten**
*name name pig*
'Naus [gave] a pig to Dege.'

Researchers claim that there is no verb 'to give' in Amele for cultural reasons. They argue that because giving is so basic to Amele culture, the language manifests a tendency to allow the 'experiential basicness' of giving to correspond to a 'more basic kind of linguistic form' – that is, zero. Nothing can be simpler than nothing after all. No verb is needed for this fundamental concept of Amele culture.

This is a good thought. In fact it is worth talking more about this concept of communicating by saying nothing. As the bluegrass singer Alison Krauss says, 'You say it best when you say nothing at all.'

We are often silent about verbs and longer ideas because our culture has added the meanings that we leave out of our sentences and stories, filling our silences with meaning. 'The living conditions were Dickensian.' What does this mean? Well, if you know what Charles Dickens wrote about, as most literate westerners do, you do not need to have it spelt out to you that the conditions probably included poor sanitation, overcrowding, and decrepitness. 'Dickensian' is a cultural cue to a body of cultural knowledge that does not need to be spoken.

The Old Testament Hebrews were prohibited from uttering the name of God. For, if you did not know the name for God, you were not a member of the culture. And if you did not know it was forbidden to pronounce the name for God, you were not a well-integrated member of the culture. The culture of the ancient Hebrews ascribed holiness to the name by silence – by declaring it ineffable. Ineffable but known.

The Wari' language, about which I co-authored the only grammar, provides profound examples of culture in its lexicon and syntax. Wari' literally means 'we.' The Wari' were, until about 1962, cannibals. They practiced two types of cannibalism, endo- and exo-cannibalism, that is, they ate their own dead (endo-cannibalism) and they also ate their enemies (exo-cannibalism). There were different rituals for each. Endo-cannibalism was a respectful, delicate affair, intended to give immortality to the deceased by absorbing them into the bodies of the living. The latter type was to show disrespect and the bodies of enemies were cut up and treated like the bodies of animals. This is all described in an ethnography of the Wari', written in 2001 by anthropologist Beth Conklin, *Consuming Grief: Compassionate Cannibalism in an Amazonian Society.*

One example of culture in the language of Wari' is the word for wife, **manaxi'** (man-a-SHE). The word for vagina in Wari' is **mana**, which literally means 'hole.' The form **manaxi'**, then, means 'our vagina' or 'our hole.' So a Wari' man will say in normal conversation something that comes out like 'My vagina and I went fishing.' There is an obvious cultural question raised by the fact that in Wari' wife and vagina are the same word. Why would the Wari' make this generalization? What does this equation mean to a Wari'? Some outsiders might jump to the facile conclusion that this is a crude and demeaning comparison, showing that Wari' men hold women in low esteem. But another possible conclusion is that we cannot know what the Wari' mean until we understand the culture that produces their meanings. Perhaps to the Wari' reproduction and the family are such important values that they honor the wife and the vagina as the source of life. So it is the highest form of flattery to call the wife 'our vagina,' the source of life. Is this a possible conclusion? Yes. Is it the right one? I don't know. No one can know unless they undertake a systematic analysis of Wari' culture, or what some anthropologists refer to as a 'thick description' of the culture.

One thing is clear, though: Wari' words are an artifact of Wari' culture. This is not surprising to linguists or anthropologists as a general principle. But it does not go far enough. While the specific example is interesting in its own right, what we really need are more examples where culture constrains grammar, as we saw in Amele's special construction for giving. Wari' provides one such example.

The most unusual syntactic construction I have ever found in more than thirty years of field research on little-studied languages is the Wari' structure for quotation. This construction is important here because, like Amele, it shows a cultural effect in the grammar – something many linguists consider impossible.

Many Amazonian Indians and other peoples around the world talk about the future, the world, and other beings by a metaphor of 'motives' and 'will.' They attribute will to the sky ('The sky says it is going to rain' is a common way of describing a cloudy sky, for example), to animals ('The tapir says it will run from me' is a frequent description of game fleeing from hunters), and to each other ('John said he was tired of talking with us' – even when John said no such thing, if that explains his behavior, then go ahead and attribute motives or will to him).

Likewise, in Wari', speakers report on others' thoughts, character, reactions, and other results of intentional states by means of quotatives. That is, they quote people as saying things, whether they said them or not, in order to communicate what the speaker believes they were thinking. But in Wari' quotatives are used for lots of other things, many more than I have seen in other Amazonian languages (with the possible exception of Kwazá, an unrelated language from the same geographical area). In the example below, which is very typical, the material quoted is in small caps. There is no verb. The quoted material acts as the verb in the sentence. And so Wari' avoids its verb 'to say' (it does have one, but rarely uses it) because of a cultural innovation:

*MA' CO*     *MAO*    *NAIN*    *GUAJARÁ$_i$*     *nanam 'o r o*
*narima$_.$*      *taramaxicon$_j$.*
'Who went to the city of Guajará?' [said] the chief to the women.'

The lack of a verb in a sentence like this is a problem for theories that separate linguistic form from culture. But this type of sentence in Wari' makes perfect sense once we recognize culture's role in shaping grammar. In Wari' the expression of the intentions and thoughts of others is fundamental to regular communication, since a high cultural value is placed on the ability to explain other people's behavior, based on their intentions. Intentions are mental facts that can neither be observed directly nor heard. Since Wari' speakers have no direct access to people's thoughts they do not use a verb such as 'to think' since this would imply first-hand information of what others think, rather than that we infer what someone thinks indirectly from their words and behavior. By 'putting words in people's mouths,' that is, attributing thoughts to them as though they were speaking their thoughts in a separate grammatical construction that does not claim that they actually said something or thought something, Wari' can express the intentions and thoughts of other people without being said to have literally attributed words or thoughts to them.

We can see more evidence of culture's influence on language in Wari' in their rich mythology. One story I like is their tale about the origin of corn. I use an English translation of this myth, rather than the Wari' original, which can be found in its entirety, along with the translation, in my grammar of Wari'. These are the first three lines of the text known by the Wari' as 'Wirin – The Origin of Corn,' told by a Wari' speaker to a New Tribes missionary in 1962:

> 'Thus was Wirin,' they [say]. 'I will clear land for a garden,' he [said].
> Garden, garden, garden, garden, garden, garden, garden.
> He finished it completely. They were all his children.

Such sentences sound alien to our non-Wari' ears. They become even stranger on closer examination. The first word in the text is 'Thus' and indicates that what is to come after proceeds logically from something else. But what is that something else? The story does not tell us. Rather, this other something is found in unspoken cultural knowledge, the 'background' of the Wari' people. It assumes that listeners to this tale are interested in hearing about foundational beliefs of the Wari' in general. And it assumes that they are interested in the origin of corn in particular. There is no verb 'to say' in this line, which is why 'say' and 'said' are in parentheses. The meaning of saying is implicit in Wari', as the verb 'to give' is in Amele.

Wari' usually omits the verb 'to say' and simply gives the content of what was said. This construction is almost unique to Wari', though in a way it is similar to examples from some speakers of English: 'So, I am like, "Dude, what IS your friggin' problem?"' There is no verb 'to say' in this English construction either, only the word 'like' used in a new jargony way.

Next in this text, the Wari' narrator repeats the word **tota** 'garden' several times. This is an onomatopoeic word as the Wari' perceive the sound of the axe in making a garden. An axe to make a garden? This sound image is based again on the background. Gardening for the Wari' entails clearing and burning a wide swath of jungle to plant corn among the ashes of the felled growth.

The line 'He finished it completely' is not unusual. He finished the chopping. There was no more to do for that garden. But the next line is not something an English speaker would utter without some sort of explanation. We suddenly learn that the gardener has been surrounded by people during the story. This was not stated previously, yet it seems to be known by the hearers. We are told that these people were his children, but have not been told that there were people there in the first place. All Wari' people know this story. They know the characters. They know the structure of a Wari' myth. So they don't need to be told that there are people there. Once again the background of culture is most clearly seen in what people do *not* say. Culture is thus found throughout discourse, in what is said and what is not said, the latter being what I call the 'dark cognitive and cultural matter' of discourse.

The principles here are easy: what we choose to say, how we say it, and what we choose not to say are all largely determined by our culture, that is, the values that we share with our community. Culture is present throughout our conversations and stories. As I have previously reported, culture also plays a major role in the sound structures of the world's languages. Values and cultural knowledge can embed themselves even more deeply into languages than the examples we just saw. One other clear example is seen in the channels of Pirahã discourse: humming, whistling, yelling, talking, and singing. I will come on to these soon, but first I want to look at another crucial point where culture and language coincide: translation.

Years ago, when I was a Christian and a missionary, I translated the gospel of Mark into Pirahã, from Koiné Greek.* That translation attempt had very little effect on the Pirahãs because it was very hard for them to understand. But this relative incomprehensibility of my translation seems to have a cultural, rather than a linguistic explanation. The lack of effect is the result of the profound cultural differences between Mark's first-century Middle Eastern culture and the Pirahãs' Amazonian culture, not to mention my slightly skewed knowledge at the time of how Pirahã grammar worked. To see the way that the translation failed the culture test, here is a brief snippet from it:

**Hiaígíai. Báakokasí higáísai, 'Ti Hisó gíxa xahoaihibiigá. Pixái ti káágai báasígi**
OK. Mark spoke. 'I want to talk about Jesus to you. Now I have pretty words.
**xahoaisoogabagaí. Hisó hi goó kaipi.**
'I want to tell you [literally: 'I pretty marks want to tell'] what Jesus did.'

Most of these words are common Pirahã words and each obeys the grammatical rules of word formation in Pirahã. And yet the Pirahãs struggled to understand my translation of Mark's gospel from this opening passage forward. I needed to know why.

The obvious and most severe stumbling block to their understanding of my translation of Mark's gospel was the nature of the information. The Pirahãs constrain what they talk about to subjects that they have first-hand knowledge of. They asked me who this Hisó was (it was meant to be 'Jesus' in this, my first attempt to transliterate the Greek transliteration (Ιησοῦς – **Yesous**) of a Hebrew name (יֵשׁוּעַ – **Yeshua**) into Pirahã). I had never seen Hisó and did not even know anyone who had seen him, so to the Pirahãs it made no sense for me to talk about him.

But there was another severe problem for the Pirahãs in trying to understand my translation: the grammatical form of the sentences I

---

* You can hear this now on the internet from Gospel Recordings at http://gospelgo.com/s/piraha.htm.

was attempting to use. The sentences above follow English structure. I knew that I had made them different from sentences found in oral Pirahã speech, but I did not worry too much about this.

Writing and speaking are different, certainly, in European languages. Few, if any, authors speak like they write. Writing is different because in it we can make our sentences longer and more complex than in speech, or much shorter and more cryptic (as with poetry). This is because the reader can take more time to understand them. Literacy is a cultural change with grammatical implications.

In my translation sample above, for example, I formed the first sentence by combining what would have been two oral sentences in Pirahã into a single written sentence. I did not know then that Pirahã grammar does not have, and nor does it allow, one sentence to be embedded into another as English does. I thought that some of the sentences I heard juxtaposed to one another were a single sentence instead of two. And, truth be told, I thought that is how they *should* say it. I had in mind that, over time, the people would adapt to my translation of Mark because that was the best way to communicate these complex and foreign thoughts in their language. I believed that, even if I was using sentences that were more complicated than those which the people used among themselves, my translation of the gospel of Mark could change Pirahã grammar, as Luther's 1534 translation of the Bible into German changed the German language. All it would have taken for my translation to do this was the founding and development of a new Pirahã group that regularly read or listened to my translation. Many languages have been changed by new literature, from idiomatic expressions to their very grammar. And very often these changes have come from translations or from reading the original languages of sacred texts, such as the Koran's effect on Arabic and the Bible's on most languages into which it has been translated.

But the Pirahãs had no interest in a translation that violated their grammar and culture by the use of subordinate clauses. Preconceived notions about grammar were partially responsible for the failure of Christian missions among the Pirahãs, as none of them, going back to the eighteenth century, learned this connection between Pirahã language and culture.

None of this is to say, however, that languages are *exclusively* products of culture. There are many components of all languages that have no cultural explanation whatsoever. And it is these other parts that produce the light 'family resemblance' that we see among the languages of the world, things that could indeed lead a Martian to think that they looked alike. This family resemblance is why extreme relativism – the idea that there are no facts common to all cultures and languages – fails. For example, all languages have nouns and verbs. All languages tell stories. All languages have hierarchical relationships – smaller items fit into larger items, such as words fitting into sentences and sentences into stories, and so on. Culture is not responsible for these things. These follow instead from our vocal apparatus, the basal ganglia, and other biological components, as well as the functions of language – including communicative efficiency, ease of understanding, informational organization, and so on.

Herbert Simon was one of the leading social scientists of the twentieth century. He won the Nobel Prize in Economics in 1978, for his research on 'the decision-making process within economic organizations,' and the National Medal of Science in 1986. Simon's curiosity was wide-ranging. His interest in decision-making and the nature of problem-solving led him to make seminal contributions to the fields of psychology, information-processing, computer science, sociology, public administration, economics, cognitive science, philosophy of science, and management, among others. With Allen Newell and Cliff Shaw, Simon invented one of the world's first computer languages, IPL (Information Processing Language), in 1956.

One of Simon's most important papers, however, turned out to be influential in all language-related fields, with the ironic exception of linguistics. In this paper, 'The Architecture of Complexity', which he published in 1962, Simon made the case that 'Hierarchy is a natural expectation, both in nature and society.' He argued that hierarchy emerges from evolutionary processes because hierarchical structures are inherently more efficient and stable than most other ways of organizing the universe. It is found in atomic structures, the organization of societies, in the way that we process information, in the organization of galaxies, in management, and in business production processes, among many,

many others. Hierarchy is usually the best solution for the organization of complex systems, a formal tool for organizing that beats all competition. We therefore *expect* to find that human languages are organized hierarchically. There is no need to call upon instincts to explain this. And languages and cultures differ so dramatically across the world that the concept of a language instinct or universal grammar is unwarranted. Simon's work on hierarchy helps us understand why we find hierarchy and common structures in all of the languages of the world.

My idea that languages and cultures interact synergistically to shape one another is not incompatible with the idea that human biology, human history, and language history place severe restrictions on the forms of languages. Hindu ascetics show us that the limits of human control over biology are less fixed than we might have thought. But they also prove that we all reach those limits eventually.

At the same time, if we are to understand actual human languages as creations of human cultures, then we must answer the question: What is culture?

For some people, the phrase 'to have culture' means that one possesses a well-developed appreciation for high attainment and beauty in art, music, literature, cuisine, relationship networks, and so on. This sense of culture is not wrong. In fact it is praiseworthy for encouraging a life of the mind. But it is prone to fall into snobbishness and is not a useful definition for the purposes of scientific inquiry.

A definition of culture we saw earlier is Edward Tylor's: 'Culture, or civilization, taken in its broad, ethnographic sense, is that complex whole which includes knowledge, belief, art, morals, law, custom, and any other capabilities and habits acquired by men as a member of society.' An alternative comes from American anthropologist Clifford Geertz, who defined culture as 'an historically transmitted pattern of meanings, embodied in symbols, a system of inherited conceptions expressed in symbolic forms by means of which men communicated, perpetuate, and develop their knowledge about and attitudes toward life.'

All of these definitions have strong and weak points. Tylor's definition served the historical purpose of focusing current thinking on a concept of culture that was in principle relevant to all societies. This is an improvement over the definition of culture as art and the 'good life,' because this latter definition is itself culture-bound and thus fails to apply to all peoples. In essence it claims that if any people lacks fine art or literature, as westerners define these, then they lack culture. Yet a weakness in Tylor's definition remains, from my perspective, that it fails to engage the concept of *values* as defining cultures. Geertz likewise omits reference both to values and to the rankings between them, things that are crucial to an understanding of culture.

All previous definitions of culture appear to share a core flaw: they treat societies' values, knowledge, and meanings as though they were fixed, when in fact they are dynamic and evolving. This dynamism emerges when we look at how cultures have changed over time.

The ability to navigate through the worlds of cognition, biology, and culture is fundamental to constructing social relationships as well, and these relationships are in turn preconditions to the development of meaning and language. Yet at the same time that we develop language, we enhance these abilities. Language is partially a thought-enhancer. Through language and culture we develop a background in common with others in our society, facilitating and deepening interpersonal ties.

One example of a language fact that could not be any other way is that our utterances are composed of parts, such as words, that come out of our mouths one at a time. If we had two mouths and two brains, we might be able to issue forth parallel streams of words, uttering more than one sentence at a time. But with our current physical architecture, words and sentences emerge one by one. We have to have words because we think in concepts and a word is the smallest linguistic unit corresponding to a concept.

Cultural effects on language can range from the trivial to the profound. Consider this scenario: two old friends meet as one returns from a year-long absence. The first one says to his returning friend, 'You have already arrived.' 'Already,' replies the friend. No one smiles. No one touches. No raised voice of greeting. No laughter. They go on about their business, taking up their normal activities. Later that evening they come together with several other men, sit around a fire, and each talks about his recent activities, people he has seen, places he has been, and so on.

Or consider this scene. John and Bill meet for lunch at a pub in

Manchester. Bill asks John what he wants to eat. John will not say 'watermelon pie.' Mancunian pubs do not serve watermelon pie. John knows this. His knowledge of his culture will keep him from making such suggestions, except to be humorous. But that is not why John does not say it. All of the knowledge that leads John to say one thing and remain silent on another is tacit, part of the background. Or suppose that John and Bill meet for lunch at a restaurant serving Brazilian food. John will not order 'steak and kidney pie' for the simple reason that such a dish does not form part of Brazilian cuisine. But though pubs and churrascarias manifest many different cultural offerings and values, they share many others. At the same time, Brazilian restaurants and Mancunian pubs have in common the cultural expectation that the customers will be served and that they will not have to fist-fight the owner for food. They might have had to do the latter, however, if they have entered a private home and ordered the homeowner to serve them. This is because western cultures, of which British and Brazilian are two examples, have in their background information the concept of 'restaurant' and buying food. But among an Amazonian tribe without much contact with the outside world, you will not find a restaurant category in the culture. You may be able to barter for food with someone, but don't take all your friends to a hut at noon expecting to buy or trade for a hot lunch.

Conventions can vary dramatically across cultures, in fact, in more ways than we tend to realize. But who are the language police? We need to ensure that speakers of languages follow their cultural values, but how do we do that? I am not talking about following the suggestions of the American English standard reference, Strunk and White, but following the right word order, not inventing your own vocabulary, and so on. What happens if an American says 'With this ring I marry you,' instead of 'With this ring I thee wed?' or what happens if a Wari' changes a beloved oral story? How is the relationship between cultures and languages made predictable and constant? The philosophical answer to this question is 'convention.'

Convention in the sense that is useful for us here is defined by Harvard philosopher Nelson Goodman as 'the artificial, the invented, the optional, as against the natural, the fundamental, the mandatory.'

This definition enables us to see convention as something that cultures invent, not merely as a general fact about the world. It is vital for us to understand that different cultures have different conventions and that these conventions may be unique, deriving from the values of a particular community.

The Scottish Enlightenment philosopher David Hume said about the relationship between convention and language that 'languages are gradually establish'd by human conventions without any explicit promise. In like manner do gold and silver become the common measures of exchange, and are esteem'd sufficient payment for what is of a hundred times their value.'

In more recent years, philosopher David Lewis of Princeton University has proposed a theory of linguistic conventions that gives some teeth to Hume's proposal. But Lewis's proposal has been severely criticized by leading philosophers and linguists for failing to take into account linguistic productivity. The main question regarding Lewis's proposal is this: If language is at least partially the result of convention, then how is it possible for speakers of languages to invent new words and utter novel sentences? If something is novel, there simply cannot be a convention about it. You cannot have prior agreement on something no one has ever heard in the first place. That is the supposed problem with convention. Among some linguists, the criticism of Lewis is based on the assumption that language emerges from properties of the human mind, not from societal constraints. But to proponents of Construction Grammar, Lewis's notion of convention fits perfectly. This is because in Construction Grammar 'novel' utterances are built from conventionalized constructions.

Yet, for any scientific proposal there are different sides, contrasting analyses, theories and predictions. Choosing between them and justifying one's choices are what being a scientist is all about. And we select our theories in part using data. For language there are sources of data that can help us evaluate the claims of nativism, though they are found in places linguists rarely look.

The culture of the Sateres, a Tupi group near Parintins, Brazil, offers some insight into the nature of convention. The late Albert Graham, my mentor in living among Amazonian peoples, was a missionary to the Sateré-Mawé Indians of Brazil for more than fifty years. When he first arrived in their territory, the Sateré asked him if he was Anumara'it, a light-skinned hero according to some Sateré legends. In Graham's understanding, Anumara'it had left the Sateré a sacred club in the shape of a canoe paddle, known as the Poranting. The Poranting had writing on both sides. Some anthropologists say that on one side were war stories while on the other was the story of how the world was created. But according to Graham, in his village along the Andirá river, it is believed that the writing on one side of the Poranting lists the good things people *should* do while the writing on the other side tells the Satere what they should *not* do. The problem is that only Anumara'it is able to read the long-forgotten symbols on the club. Until he comes again, no one can know how to distinguish the good from the bad on the Poranting.

There really is a Poranting. And it really is the case that, to date, no one has deciphered it. Some in Sateré society originally believed that Graham might have been the one sent by their god, Tupana, to give them back their moral code. But this puzzle at the heart of Sateré-Mawé culture remains unsolved.

The immediate reason that the Sateré-Mawé can no longer understand their own Poranting is simple: they have forgotten the social agreements that may once have existed for linking the form with the meaning of each of the signs on the Poranting. These agreements or conventions on how to extract meaning from the world around us, whether from an ancient club, hand gestures, speech sounds, or stop signs, form our foundational social contract, upon which all other contracts are built in all human societies.

Why, for example, in English does 'pen' mean 'ink-dispensing writing instrument'? Why does 'arm' mean 'one of two major appendages of the upper human body'? Words mean what they do because they are signs determined by convention. We, or rather the speakers who first used a word or phrase, agree that the sign – word, gesture, or sound – means one thing and not another and that is has a particular form. We promise consistently to use the word in question to evoke this meaning

In this sense the origin of words and other linguistic forms is similar to the origin of jokes. We know that somewhere somehow there was a first teller of a particular joke. We will probably never know who that was. And it really doesn't matter, so long as the joke keeps circulating. It may have originated with the original teller as an individual invention, but it only can be said to begin life as an artifact of the general culture after it has spread. And no one has a better explanation of what causes jokes, words, or pronunciations to spread than 'because people like it,' because it hits a cultural nerve. A funky new word spreads for the same reason that a dress style does.

Cultures can alter their contracts as they wish, however. For instance, the word 'bad' for all English speakers means that the object it describes is viewed negatively by the one uttering it. But it could mean the opposite if we agreed to make it so. That is in fact how many speakers use it in sentences like 'Dude, that Corvette is *bad*.' Or we can alter its meaning in a different direction to mean 'tough,' 'not to be messed with,' as in Jimmy Dean's song 'Big Bad John.'

We agree on our languages in the same way that the Sateré-Mawé agreed on the Poranting. And our agreements go beyond words. This story might sound strange to some, though. You might retort, 'This doesn't make any sense. I never agreed to any of this stuff. Words just mean what they mean!' I need to clarify. When I say that our languages, words, and grammars are contracts based on (thousands of) agreements and sub-agreements, I do not mean that our agreement is conscious. And many of the agreements we now live by, like the US constitution, were made by people who died long ago. But we must 'go along' with this language contract, assenting to language conventions by our use of them, or we are doomed to be language-less.

Agreements are found in all languages everywhere. For example, Banawá is one of a small family of languages known as Arawá, whose range stretches from Peru to the state of Amazonas in Brazil, and all of whose speakers live near the deep, muddy water of the Purus, one of the world's largest rivers. The Banawás live along a small tributary of the Purus, the Kitiá. They speak a dialect (as American and British English are dialects of the larger language 'English') of the Madi language, whose other two dialects are Yarawara and Yamamadi. There are only eighty or so Banawá living and most of them speak Portuguese almost as much as they speak Banawá. Nearly half of them now live in Brazilian towns and smaller communities along or near the Purus.

I studied the Banawá language on and off for several years, mainly in order to help other linguists make headway in the language. But I did take time to live among the Banawá people for about three months, working hard to analyze their word structures, their sentences, phrases, and stories. Like Spanish, German, and many other languages, Banawá and its sister languages classify all nouns for gender, either masculine or feminine. But there is a twist. In Spanish and many other languages, the 'default' gender is masculine. So if you don't know what something is, you refer to it as masculine. If you see a mixed group, both masculine and feminine, you refer to it as masculine. Only if a group is all female do you refer to it as *ellas*. But if just one male joins that group, it is *ellos*, masculine. If you put a new word into Spanish from another language, one whose gender you are not sure of, then you classify the word as maculine. English does not actually have gender marking outside of the pronouns 'he' (masculine), 'she' (feminine), and 'it' (neuter). But if we are referring to a person or thing and we do not know its gender, or we are referring to people generically, we use the form 'he' for the most part – masculine is default for English, too.

But in Banawá, feminine is the default gender. In fact, feminine is not only the gender for new words and mixed groups, it is also the gender that all Banawá speakers use for the first and second person. That means that, regardless of whether a man or a woman is speaking, they will refer to themselves as feminine. If I say 'I did this' or 'I did that,' I will use the feminine gender, because the pronouns for 'I', 'me', and 'you' in Banawá are always feminine. Since the verb often agrees in gender with the subject in Banawá, that means that when I, a man, am speaking, both the pronoun and the verb indicate feminine gender. In all the Arawan languages, in fact, this is how gender works. Here is an example from Jamamadi (zha-ma-ma-GEE):

**Kodzo tsʰite onaharo**.
*'I [feminine] shot the lizard [masculine].'*

**Kodzo tsʰite onahari.**
*'The lizard was shot by me.'*

These two examples are exactly alike until the last vowel. The second one ends in the feminine i and the first one in the masculine o. The first example corresponds roughly to an English sentence like 'John saw Bill.' The second one corresponds to a passive like 'Bill was seen by John.' How does a speaker tell the difference? By the gender – the o v. the i at the end of the verb. In the first one, the person doing the shooting, me, is the subject, so the verb agrees with it, and is feminine. In the second, the lizard is the subject, so the verb agrees with it – but lizards are masculine in this language.

The use of gender to mark subjects in the Arawan languages teaches us a great deal about the nature of grammar. The natural question to ask, then, is whether there is a link between Arawan culture and gender marking. The only answer at present is, 'Perhaps.' I do not yet have a solid case to make. What *is* clear is that cultural relationships between the sexes in Arawan cultures are also different from those of many other groups. Generally, people of the opposite sex do not speak directly to one another unless one is pre-pubescent, they are siblings, or they are married or in a serious relationship, or in certain special circumstances, such as when I, the visiting linguist, ask them questions. When I lived among the Banawás, the men would come over to visit as a group. After thirty minutes or so, they would leave and the women would come to visit my wife, while I left the room. Segregation of the sexes is a highly ranked value in Banawá cultural life.

Distinctions of gender are further underscored in Banawá culture by their puberty rites for women. During my stay among the Banawás, several girls were kept in huts, but I was not there when any of them were released. What happens to them was described to me by Bido Banawá, headman of the village.

Not long before a young girl menstruates for the first time, her father builds a small hut that has only a single opening. The hut is placed behind the father's house. The single opening is only large enough for someone to crawl through. As soon as the young girl menstruates, she moves into the hut. The Arawan cultures believe – or at least the Banawá, Jamamadi, and Jarawara do – that if the young girl looks on a man, he will develop severe stomach problems, perhaps leading to his death. The girls are allowed out to relieve themselves and to bathe, but they must always wear a basket over their heads, woven so tightly that they cannot see through it, and they are required to be led around by female relatives.

After several months, the father and the village headman meet to decide when the girl can come out of her '*corral*,' as the men jokingly refer to it in Portuguese. A huge party is planned. The men go hunting for days and smoke, dry, and salt meat for the party. Women bring in large quantities of legumes from their gardens. When everyone is ready, the girl is released from her confinement and taken to a cleared space in the center of the village. Most of the night there is singing and dancing, women and men separately. The girls is led by the other women around and around in a large circle during this time. At the end of this walking, dancing, and singing, all of the women strip the clothes off the girl and hold her face down across a raised platform. The girl is then beaten by all the men with sticks until the sticks break or the men decide to stop. The girl is then released, with a bloody back, often unable to walk for a day or more, according to Bido. Once this beating is over, the girl is ready for marriage.

It is exactly by exploring such cultural values that we would try to build a connection between feminine identity and grammar in Banawá and other Arawan languages. I have not yet established such a link, but I am working on this. One of the hardest things about establishing such causal relationships between culture and grammar, however, is that language and culture often evolve along divergent paths for historical reasons. Many cultures of the world, for example, are moving towards more western styles of dress, commerce, and so on, while their languages, apart from vocabulary, remain relatively stable.

This stability is not inviolable, however, since not only loan words, but also new types of constructions often enter languages through cultural change. In English, for example, we say, to take just one example, 'If you please...' – an unusual word ordering for English that is used in more formal speaking. This phrase was inherited directly from the French '*s'il vous plaît.*' A phrase that is borrowed and translated, while maintaining the grammar or word order of the original, is called a 'calque.' It is very hard to establish that the culture has affected the grammar in this way, but it can be done. What we have in the case of the

Banawá is an unusual cultural system of gender relations in combination with a very unusual grammatical gender system. That should raise an eyebrow and lead to further research.

Frequently, as cultures come into contact with one another, one culture's lexical conventions are used to fill gaps in another community's vocabulary. In these cases, we say that one community borrowed the word from another. Do you eat or at least use the word 'haggis'? Then you borrowed this word from Scots Gaelic, where haggis originates. Just as many cultures have borrowed English expressions and words like 'OK,' 'know-how,' and 'pancake,' we English speakers have borrowed many words from French and Latin. Some estimate that as much as 50 per cent of modern English vocabulary comes from French or Latin. A few examples of such borrowings just from French are: baguette, judgment, litigation, azure, garage, ennui, village, cigarette, coup, clique, critique, limousine, maneuver, mayonnaise, vandalism, voyeurism, panache, detente, and espionage, among many, many others.

In trying to connect cultural values and language, we need to exercise considerable caution, however, because languages and cultures do not always match up straightforwardly. This is illustrated by two cultural regions of regular tribal contact, the Xingu Park of central Brazil and the Vaupés region of northern Brazil.

The Xingu (sheen-GOO) Park, located in the state of Mato Grosso, Brazil, is over 26,000 square kilometers in area and was created on April 14, 1961 under the administration of President Jânio Quadros. There are several cities nearby, including Canarana, Paranatinga, São Félix do Araguaia, São José do Xingu, Querência, and others. The tribes located in the Xingu include (with their estimated populations in parentheses): the Kamayurá (355), Kaiabi (745), Yudjá (248), Aweti (138), Mehinako (199), Wauja (321), Yawalapiti (208), Ikpeng (319), Kalapalo (417), Kuikuro (415), Matipu (119), Nahukwá (105), Kĩsedje (334), and Trumai (120). I have visited the park in my research on the language of the Kĩsedje people.

In earlier years, the Xingu area was one of the places that American and European adventurers, such as Percy Harrison Fawcett, whose ignorant wanderings are described in David Grann's 2009 book *The Lost City of Z*, searched for hidden riches. My former student Michael

Heckenberger of the University of Florida's Department of Anthropology has become well known for his pioneering archaeological research in this area, discovering the existence of pre-Columbian settlements far larger and more complex than anything anyone had previously estimated.

What interests me about the Xingu, though, is that the cultures of the fourteen groups currently residing in the park are very similar, yet their languages are very diverse, representing multiple linguistic families. Despite this, the peoples of the Xingu participate in cultural ceremonies together, work together for land rights and political representation, and engage in a range of other activities together. What this all shows us is that cultures may change, or intermingle, but languages remain stable. Therefore, when a link between culture and language is proposed, it is vital to determine whether the connection is the result of the current state of the language and the culture or whether it is a vestige of a previous period, but evidence for the link has been eliminated. In other words, it may be that some cultural influences cannot be traced and that the facts of language that the culture might have produced at a different point in time look arbitrary from a contemporary perspective.

Traveling over one thousand miles north of the Xingu you reach the Vaupés river, where there is a people, the Tucano, who value 'linguistic exogamy' – the practice of a man marrying a wife who speaks a different language from himself. Tucano villages are located in both Colombia and Brazil, although most are found on the Colombian side of the border. It is normal for Tucanos to speak two, three, or more languages, and any Tucano household is likely to be host to numerous languages. The languages represented in these communities include the Bara Tukano, Barasana, Desana, Macuna, Wanano, and Tucano.

The Tucanos are thus a relatively homogenous cultural group that *insists* on linguistic diversity. This does not at all mean that there will not be connections between culture and language in Tucano – to the contrary, in fact – but it does mean that care will be needed in drawing any such conclusions.

Because they are relatively free from further interventions from western cultures, areas like the Xingu and the Vaupés actually give us a great chance to test the language–culture connection. We would expect each of the languages at some point to take on similar characteristics deriving from their similar culture. No one to date has studied this. It is an exciting research possibility.

We have talked so far about a number of ways that culture affects grammar, from words to syntactic constructions to translation. Another important connection is story-telling. This cultural effect can often be seen best when the story-telling takes place across cultures and the values of one of the cultures conflict with the values of the other.

In past years, Kayapó leaders Paulo Paiacan and Raoni were two of the best-known and most articulate spokesmen for the tribes of Brazil's national Xingu Park. Paulo spoke fluent Portuguese and constructed his sentences and arguments to appeal to educated Brazilians in positions of power. Raoni was one of the survivors of his people's first known contact with Brazilians in the 1940s. He was tough as nails, had killed many men with his war club and arrows, and wore a plate in his lower lip, as many of his older Kayapó kinsmen did. Raoni spoke from the heart, in imperfect Portuguese, with pride and charisma. The Kayapó benefited greatly from effective speakers like Paulo Paiacan and Raoni, at least as much as from their colorful body paints, feathers, and formidable physiques.*

---

* Paiacan emerged in the past couple of decades as a worldwide symbol of the Brazilian Indians' defense of their Amazon homelands. He was called 'a new Gandhi' by Anita Roddick, founder of the the Body Shop. *Parade* magazine called him 'A man who would save the world' in a cover profile. People wanted to make a hero, the ideal Indian leader, out of him. Like other Kayapó, he was powerfully built and had long, black hair. He had won many international awards, appeared on US television, and was even a character in some children's cartoons. Director Ridley Scott apparently planned to make a Hollywood movie about his life. But Paiacan was a rapist. He was found guilty of raping and trying to kill an eighteen-year-old woman. This incident was used for racist purposes in two ways. Some said that this just proved that all Indians were savages. Others said that it proved that all white men were savages because 'Indians don't do this' (that is simply silly, I should say) and that Paiacan must have learned his aggression from *brancos*. Even one of my anthropological heroes, Darcy Ribeiro, a former Brazilian intellectual leader, got into the fray, exclaiming emotionally rather than rationally that 'If a rape occurred, it was a reflection of his living among white people.'

Kĩsedjê leader Kuiussi is another proud and formidable indigenous leader of Brazil. He is conscious of his place in Brazilian history and of his role as a leader of people of the Xingu Park, as well as his position as the main chief of the Kĩsedjê. He had told me before we traveled together in Europe that

> There is a statue of me in Peru, because I am a leader of indigenous peoples of South America. I have been many places – to Brasilia, Rio de Janeiro, São Paulo, and Washington D.C. But traveling, seeing pretty buildings, having statues made of me – none of this means anything to me. I do not like to travel. I am getting old. I prefer my hammock in the village with my children and grandchildren. I travel to help my people. When I get back from travel, my people want more from me than the promises of white men. They want things they can use – guns, motors, seeds, tools, and things to help our lives. I must show them that we have help to protect our land, to fight off the farmers who destroy our rivers, with pesticides from their farming. I must think of the well-being of my grandchildren.

Then, as he stared at nothing in particular, he concluded, 'It is not easy to be a chief. It is a big responsibility.'

On May 3, 2005 I sat on the stage of the main lecture hall of the School of Oriental and African Studies in London with Kuiussi and two other representatives of the Kĩsedjê. The three of them had painted themselves in bright urucum red and charcoal black and donned radiant halo-like headdresses made of homespun cotton resplendent with feathers of macaws, toucans, and parrots. Attention was drawn to their powerful biceps by tightly tied, white cotton armbands, brightened by orange, yellow, black and red feathers. There were rattling bands of Brazil nut shells tied about their ankles, to beat a rhythm as they danced.

Kuiussi, his son-in-law, and his brother had just danced and sang

---

The answer is simpler. Some aspects of human biology – lust, drunkenness, and violence – override cultural values sometimes. No culture values rape. No culture values violence. But all have them.

in different Kĩsedje styles. The Kĩsedjes are tall, like the Kayapó, their closest relatives culturally, genetically, and linguistically. The men average nearly six feet and 200 pounds. They have earned the respect of all other Xingu tribes, even of the Kayapó. They smile and laugh little around strangers and they look all business in their dancing paint. It didn't surprise me, then, that they showed no fear or nervousness singing and speaking to about 400 people at one of Britain's most respected institutions of higher learning.

Kuiussi approached the microphone, war club in hand, and began to speak. I had no idea what he was going to say. I had organized and gotten funding for this trip so that the Kĩsedje could appeal for financial help for them to restore land recently returned to them by the Brazilian government from local agrobusinesses – land nearly ruined by pesticide-intensive planting and cattle packing down the soil so hard that it had become unusable for traditional agricultural methods.

Months before, I had also offered to help Kuiussi with his speech, to which he replied firmly: 'My words were given to me by my people and I am not speaking for a white man. I want no one to think that you are telling me what to say.' So I was looking forward to his speech, knowing little about its content other than that it would be accompanied by colorful photographic slides of the people taken by government employees, anthropologists, and professional photographers over the years.

Kuiussi knew, as any good speaker knows intuitively, that the content of words must be showcased in a form fitting and effective for the transmission of the message at hand. He recognized that his language tonight was his most important and most appropriate tool to help his people.

His eyes took in the entire crowd. When he began his address with 'I do not know if you have the courage to listen to me …' I knew that this was going to be an important event. He spoke to the audience as *brancos*, seeing them as the very people who had exploited his people and all indigenous communities of the Americas. Kuiussi spoke about what whites had done to Kĩsedje land and remarked on the irony of the fact that to preserve his own culture he needed the support and tools of the European cultures that had ruined his people's hunting grounds, polluted their river, and reduced their arable land to a tiny fraction of the original amount they had farmed on.

But many in the crowd were shocked by Kuiussi's attitude toward them. They apparently had expected a sweet, loveable old tribal leader to lament the loss of his ancient ways and to tell his audience about all that was being destroyed in his culture and language. The audience might have also imagined that Kuiussi would ask them for contributions to help his children. Instead, they got a crusty old warrior telling them that they had a *responsibility* to give to repair for the activities of their fellow Europeans and that he would tell them what they should give for.

When he got around to demanding a tractor, I could see that some people had nearly stopped listening. I could hear some of them whispering words to the effect that 'That's a western device! Who does this guy think he is asking for something like that?' They refused to hear Kuiussi explain that the tractor was to till the land that had been packed down to uselessness by herds of cattle that ranchers had turned loose on it. That it would help them build barriers on river banks to help prevent run-off of pesticides from surrounding Brazilian fields.

Some people in the audience took exception to the speech. But Kuiussi believed he had said what it was his duty to say. He felt a responsibility to bring back results to his people. He was not accustomed to compromising when speaking for his people. Yet his audience that evening was not used to feeling morally upbraided just because they had come to a talk such as this. They did not want to be preached to. It just was not done.

Although many people did connect with Kuiussi's message and although Kuiussi did appreciate that gifts from others required a deeper and more generous spirit than many whites he had dealt with in the past, the Kĩsedjê ultimately found their visit to Europe unsatisfying. People looked at the Kĩsedjê and admired them but did not offer any significant financial help.

Kuiussi spoke as a chief, not as a supplicant. Nor as a curiosity. In a cross-cultural encounter of this sort, communication often breaks down, just as it had this evening. It was not because of poor translation or because the hearers and speakers used language differently. British and Kĩsedjê cultures share important uses and conditions for language.

But the bold challenge Kuiussi issued backfired on him, in spite of the integrity of his intentions and preparations. Kĩsedjê culture clashed with European culture in the use of language. The conflict was between his community-given content, his chiefly, warrior delivery, and the expectations of his audience.

Like most peoples, the Kĩsêdjê associate certain functions of language, such as speaking on behalf of the people as a whole, with political and cultural roles under the exclusive purview of their chief or his designee. This is not unusual. Europeans, Americans and many other societies have similar constraints. Kĩsêdjê and American audiences expect their leaders to adjust the form of their speech to the relative gravity of their purpose for speaking.

Anthropologists particularly concerned with the ways in which societal and cultural values can affect the use of language are known as linguistic anthropologists (or, depending on the emphasis of their studies, anthropological linguists). Linguistic anthropologists are often concerned with how power and language interact in specific cultures. So they ask questions like, 'If only Kuiussi or his designate may speak in the name of their community, how do they come by this authority? If his speech has power, where does the power come from?'

What is most important in understanding speeches like Kuiussi's is the origin of such oratory in the comprehension of language more generally and how communication serves people in their daily lives.

Early anthropologists often took it as obvious that some cultures were superior to others. To some of these eurocentric folk, a jungle community without written literature, without doctors, lawyers, professional philosophers, composers and the like is clearly inferior to European civilization with its Mozart, Isambard Kingdom Brunel, Albert Camus, and their ilk.

But these views are, ironically, anti-Darwinian. I say 'ironically' because this kind of thinking originated as so-called 'social Darwinism.' This was, in its most extreme form, a justification for European imperialism. The basic idea was that societies advance in stages from primitiveness to civilization. Europeans have reached the highest stage of development in this view (what a coincidence that the propounders of this view saw themselves at the apex of cultural evolution). Simpler societies are in the lower stages of development by this perspective. But if we took a Darwinian view of cultural differences, we would see cultures as developing to fit particular environmental needs. An Amazonian culture might lack classical composers, for example, principally due to its size, history, and environmental needs. Certainly not because the people are inferior. For the sake of discussion, Mozart may have been the smartest human to have ever walked upon our planet. But this no more makes Austrian culture the highest on any cultural evolutionary scale than Hitler's birth in Austria makes Austrian culture the worst. And Mozart and Hitler do not somehow cancel each other out in a misguided ranking of Austria on some grand culture scale. Each is a product of personal attributes and personal history, working together with cultural nurturing, like language.

Amazonian or New Guinean or African or Australian or other societies without writing might have produced the greatest explorer, or the greatest survivalist, or even the greatest thinker in world history. But because they had no pressing need for – and hence did not develop – a writing system, we will never know. And it is pointless to speculate because the answer to such 'great man' views of history is of marginal interest to understanding human beings as a species. Tolstoy made this very point brilliantly at the end of *War and Peace*: 'In historical events great men – so-called – are but labels serving to give a name to the event, and like labels they have the least possible connection with the event itself.'

Some philosophers have proposed that all social institutions, such as money, the presidency of the United States, universities, legal contracts, and so on, are fundamentally linguistic. They do not mean by this simply that we have to be able to communicate in order to talk about institutions, reach agreements, and the like. They mean something much stronger – that our societal institutions are made from language, and that they could not exist without declarations like 'This is a court,' 'It is illegal to jay-walk,' 'This piece of paper is legal tender,' and so on.

So we have seen that most of human language, its forms and functions, result from the interactional or social instinct proposed by Aristotle in conjunction with various features of the real world.

Language is the tool by which we created our social world. But as we use a tool, we modify it and shape it to serve us more effectively. Language has been shaped in its very foundation by our socio-cultural needs.

With this understanding of aspects of the origin of language, we need to explore in more detail how the use of language establishes its status as a cultural tool.

## Chapter Nine

# LANGUAGE THE TOOL

*'But what is the meaning of the number "five"? No such thing was in question here, only how the word "five" was used.'*
Ludwig Wittgenstein, *Philosophical Investigations* (1953)

There is a well-worn 'metajoke,' a joke about jokes, that I like. It goes like this: 'A guy gets sent to prison. He is nervous and wants to fit in. On his first day, in the dining room, he watches one of the hardest-looking criminals shout out "Two!" Everyone (except the new prisoner) breaks out in hysterical laughter. Another convict yells, barely containing his laughter, "Six!" All the prisoners laugh, some rolling on the floor, holding their bellies.

'The new arrival asks his cellmate that night, "What the hell is so funny about numbers? Why did people laugh just because someone yelled out a number?" His buddy answers that they have been in the big house for so long that they all know the jokes, so they just yell out the numbers and people remember them and laugh. "Oh, I see," says the novice, determined to learn the numbers of all the prison jokes.

'After a year, he has mastered the system and works up the courage to participate. Nervously he announces loudly "Seven!" Nothing, *nada*, not a single laugh. He tries again, with a number that always gets a laugh, "Two!" Again, nothing.

'That night in his cell he asks his bunkmate about it. "Why didn't anyone laugh when I called out joke numbers today?"

'"Well," his friend condescends, "some people can tell 'em, and others can't."'

At the risk of spoiling what little humor this joke might carry, let us look at it anthropologically. In anthropological terms, even if two people utter the same thing, that does not guarantee that their identical utterances will produce the same effects in their listeners – cultural values applied to individual situations are crucial in determining how we react and interpret the sounds, sights, and creatures around us. For example, if I tell you that 'John is so smart!' you could think by that that I mean he is stupid, the opposite of what I intended, depending on your expectations of sarcasm *v.* literalness in your standard conversations. Our conversations with our cultural peers reveal more about our values, knowledge, and expectations than any other form of linguistic behavior.

That is why the essence of language as a tool is seen clearly in leisurely conversation. I recall a pleasant evening years ago when the Christian linguist Kenneth Pike offered the invocation for a large dinner of linguists and anthropologists by praying, 'God, thanks for good food, good friends, and good things to talk about.' I agree with him completely. These are three of the most wonderful experiences of being human, especially the communion and communication. But why? What goes on when we talk? How do we communicate and what do we communicate? Are the answers to each of these questions different for different peoples or are they the same across cultures and societies? This is the crux of the case that needs to be made about the deep connection between language and culture.

If it makes sense to explain human language as a cultural tool, then part of that explanation must include what it is a tool for. Is it a tool to communicate or a tool for helping us to better formulate and express our thoughts? Some linguists have been telling us for years that language is not a tool for communication. They reason that it is not as well suited to communication as we would expect if that were its main purpose. The evidence adduced to support this claim comes from language phenomena in which communication is poor – 'garden path sentences,' ambiguity, vagueness, and a few others, all of which can adversely affect clarity of information transmission.

These phenomena are natural effects of language design and some subset of them is likely present in one form or another in all languages. Because they all create problems of one sort or another for communication, so the story goes, they are *prima facie* evidence that the primary function of language is not communication.

But there are two broad reasons to reject this view. First, it has conceptual problems. Second, it has practical problems. Conceptually, this view confuses problems in *planning* to speak with the inherent structure of language. Here are two kinds of sentences that are often used to claim that language cannot be for communication. The first set illustrates 'garden path sentences' and the second ambiguous sentences. Both are illustrated below with examples from Wikipedia:

*Garden path sentences:*
The old man the boat.
The man whistling tunes pianos.
The cotton clothing is made of grows in Mississippi.
The complex houses married and single soldiers and their families.
The man returned to his house was happy.
The government plans to raise taxes were defeated.

*Ambiguous sentences:*
Prostitutes appeal to Pope.
Complaints about NBA referees growing ugly.
Hospitals are sued by 7 foot doctors.

Let us take garden path sentences first. These are hard to interpret, though they are frequent in written language (newspaper headlines come to mind). Their problem is that they 'lead us down the garden path' of interpretation. That is, they have a different interpretation at the end than the one we thought they would have when we began to read or hear them. Perhaps the most famous example of this type of sentence is 'The horse raced past the barn fell.' Most people read this sentence without problem until the last word. Then, as many experiments have shown, they start all over again, sometimes more than once, until they get the right interpretation, which is 'The horse that was raced past the barn fell,' which someone might use as an answer to the question, 'Which horse fell?' 'Why, it was that horse that the stupid jockey raced past the barn.' What makes this sentence (and all other garden path sentences) hard to understand is the omission of words that, though not required, help with the interpretation. In English, for example, when we utter a relative clause, we can choose to include or omit the relative pronoun and the verb: 'The horse *that was* raced past the barn fell' *v.* 'The *horse* raced past the barn fell.' Or 'The man *that was* here yesterday is here again today' *v.* 'The man here yesterday is here again today.' Anytime we omit words, we risk causing our hearer to misunderstand us. Leaving out the relative pronoun and the verb 'to be' (as when we left out 'that was') can cause a 'mis-parse,' a false analysis of the sentence. Linguists recognize that this type of sentence is a natural by-product of human grammars. Interestingly, though, these sentences do not confuse us when we think them to ourselves. If I just happened to think the thought 'The horse raced past the barn fell,' it would not confuse me. *I* know what I am thinking. But such sentences often do confuse hearers and readers. Therefore, some conclude that language likely evolved primarily for the purpose of expressing thoughts and that it is only secondarily useful for communication.

One immediate rebuttal to this theory is that such 'problems' are largely artifacts of reading, rather than speaking. This is because in speaking and hearing we have the benefit of intonation, the pitch patterns of sentences that chunk the sentence into its main parts – to help the hearer interpret what we are saying. Intonation helps keep spoken language clearer in meaning than other forms of language. In thinking, they are also not a problem because *we* know what *we* are thinking. We do not form sentences that are ambiguous, vague, or garden paths to ourselves.

The same point that some researchers make about garden path sentences could also be applied to other 'deficiencies' of language such as ambiguity and vagueness. On the one hand, we should realize that for all of us there are many times when it is useful to communicate ambiguously or vaguely (ask any politician). These are useful tools, not design defects, and they can play an important role in communication because not all speakers want their communication to be clear. We do not communicate solely to transmit information but to produce an effect – a behavior or a way of thinking – in our hearers. To better understand these supposed problems, a more careful consideration of what they are might help.

A vague sentence is a sentence with a poorly specified meaning. So, 'He won't be here today' is vague because, without more information,

'He' could refer to any male on the planet. Unless we can narrow down the meaning with additional context, we cannot figure it out. But this does not happen when we think 'He won't be here today', when we know exactly who we are referring to. Does this mean that vagueness is evidence that language is for thinking first and communication second? No, actually. What it means is that it is *useful* to be vague with others on occasion. It is not useful or even possible to be vague with ourselves. From this perspective, vagueness is a design advantage for a system built to communicate.

Ambiguity is also found in words and in sentences and is the property of possessing multiple definitions. A word like 'bank,' for example, can refer to either a financial institution or the side of a river. Ambiguity in sentences is seen in the famous example 'Visiting relatives can be a nuisance.' This is ambiguous because it has two meanings: 'relatives who come to visit me can themselves be a nuisance' and 'it is a nuisance for me when I am obliged to visit my relatives.' But like vagueness and garden path sentences, ambiguity does not seem to create a problem for us when we are thinking. In fact, ambiguity is a natural by-product of language as a communication system.

Recent psycholinguistic research on ambiguity concludes, in fact, that ambiguity is a desirable and expected feature of languages if the function of human language is, as I have been arguing, communication. As Steven T. Piantadosi and colleagues put it in their 2010 paper 'The Communicative Function of Ambiguity in Language:'

> We argue, contrary to the Chomskyan view, that ambiguity is in fact a desirable property of communication systems, precisely because it allows for a communication system which is 'short and simple.' We argue for two beneficial properties of ambiguity: first, where context is informative about meaning, unambiguous language is partly redundant with the context and therefore inefficient; and second, ambiguity allows the re-use of words and sounds which are more easily produced or understood.

Although it sounds complicated, what this means is straightforward. We could avoid all ambiguity in languages if we simply learned different forms for all sentences and all words. But this would create a huge list of things for us to learn. As the authors put it, it is efficient to reuse words in multiple contexts. In English, there are many ambiguous words in spoken speech. Ignoring the way that words are written for now, think of homonyms like 'too,' 'to,' and 'two.' If I just say these words out of context, a native English speaker will not know which one of them I intend. But if I were to say, 'I want *to* go,' all English speakers know that I do not mean 'too' or 'two.' Likewise if I were to say, 'I want *two* of those,' no speaker of English would think I meant 'to' or 'too.' Being able to reuse these forms makes the total list of forms in the language smaller and thus more efficient. The space between my mouth and your ear is an information channel when I speak. If every word is different, the information rate is very high and the speech channel will have the limits of its bandwidth tested. But if many words sound the same, but are distinguishable by context, then the communication system overall is more efficient. This study, again from researchers at MIT's Brain and Cognitive Sciences department, demonstrates perhaps more clearly than any study for years that the function of language is not for the expression of thought but for the social purpose of communication, as Aristotle recognized over two thousand years ago.

Even without this important study, however, it is no more convincing to argue that garden path sentences, ambiguity, and vagueness are evidence that language is not for communication than it is to say that running is for exercising on the spot rather than for locomotion, based on the greater probability that we might trip over something if we actually run. This is the practical problem of claiming that language is not for communication. As we have seen, both ambiguity and vagueness can actually be useful for a person's communicational goals – they want to say something, but they *do not want* to be precise. But if I utter a garden path sentence or an ambiguous or vague sentence unintentionally, this still does not prove that language is not primarily for communication.

It would no more imply this than the fact that I might trip while running demonstrates that running is not for moving about. When we engage the world with any human tool, cultural or biological, we invite the world to engage with us as well. We minimize adverse engagement with the world when we *plan* our activities, based on the data from our surroundings or past experiences. If you're going to run, don't run where there is oil on the floor. If you're going to talk, think first.

Whenever we utter anything to anyone, we accept the risk that those we speak to might not understand us. We are each different people, after all, with different experiences, different levels of interest in talking, different vocabularies, and so on. I must plan ahead, therefore, before I start talking or writing. As the Enlightenment philosopher John Locke put it, we must 'understand before we speak.' If I create and use an unclear sentence when I intended to communicate clearly then this is *my* fault, not my language's fault. I just didn't plan my discourse well. If I run into a stationary chair, I just didn't plan my run well.

We should never forget that language, like the rest of our cognition and physiology, is at best an imperfect solution to the common problems faced by our species. After all, people divorce because they do not understand one another. Sermons are preached about what the Scriptures 'really' mean – because the Scriptures often communicate poorly with a modern audience far removed from the original time and circumstances in which they were written. Arguments can be pointless, neither side understanding the other. Wars can begin because of poor communication between leaders. Court records show that witnesses and attorneys, judges and juries misunderstand one another frequently – even when they are supposedly speaking the same language. So language fails for communication even where it is under strong pressure not to. Language is a leaky bucket. But then everything that lives or that is designed by something that lives is imperfect and language can never be better than the thought that has gone into planning to use it.

So what do any of these failures have to tell us about the central purpose for which language evolved? Nothing much. It isn't walking that fails the clumsy toddler or the drunken lush, but they who fail at walking.

So, OK, language breaks down from time to time. It isn't perfect. And it often unintentionally confounds interlocutors – obscuring rather than facilitating communication. Even when they are used in grammatical sentences and pronounced clearly, using garden path sentences, ambiguity, vagueness, and so on inadvertently will not make anyone a better communicator, though if they are used purposely, their effect can be to enhance communication, or at least the recognition of the skill of the communicator. Speaking or writing bad sentences is something we all do every day. But these cannot be blamed on language – try telling your editor or spouse that 'I'd like to say this more clearly but this god-damned language won't let me!' Doesn't work. Like bad planning in walking, designing a building, or answering a math question, the blame for error rests on the speaker, designer, or test-taker, not on walking, architecture, or math. Shovels aren't perfect either. Not even for digging. You can take a toe off with one.

Other writers have pioneered the idea that language is a tool. Lev Vygotsky was one of the first to develop a coherent picture of the 'language is a tool' hypothesis. A Soviet psychologist and writer who died in 1934, aged only thirty-eight, Vygotsky, worked for many years in deep obscurity, but his contributions are now recognized as seminal. He believed in the Marxist proposal that changes in society and material life produce changes in human nature.

He therefore reasoned that adult intellectual functions, including language, were formed largely by social processes, growing out of an interplay of social values, cultural meanings, and the use of language in everyday contexts. As Vygotsky put it, 'Like tool systems, sign systems (language, writing, number systems) are created by societies over the course of human history and change with the form of society and the level of its cultural development.'

The proposal that language is a tool for community and communicating is insightful and pregnant with implications for an understanding of the origins, use, and nature of language. Although all researchers recognize that language is utilitarian, few if any place this utility at the center of their explanations.

Europeans and other more commonly multilingual populations often have an easier time appreciating the utilitarian nature of language than monolingual populations such as most Americans. It is not surprising that Europeans have always seen differences between languages as less of a big deal than Americans. This does not deny for a minute the political and emotional effects of language divisions that come between different ethnic groups in Europe, but Europeans realize that they live in a multilingual world. They are surrounded by different languages. The stereotypical educated European will speak at least a couple of those.

But the Europeans' mastery of other languages is not merely the result of the fact that nearly all the languages of Europe (except Finnish, Basque and Hungarian) share a common ancestor – Indo-European, a language that was spoken more than six thousand years ago and from which English, German, French, Spanish, Italian, Romanian, Sanskrit, Latin and many other languages derive. The reason is more utilitarian, motivated by politics and economics and less to do with language history. Many Americans, like other populations, live separated from other languages by economic, social, and geographical barriers. Europeans, on the other hand, come from smaller countries with relatively greater contact with other societies and languages. They need to communicate outside the bounds of their native languages. So they form a community that extends beyond the traditional boundaries of language.

It used to be easier to communicate in Europe, of course, when nearly all spoke versions of Indo-European. But as communities of Indo-European speakers became separated from one another, their tongues began to differ, their languages began to change.

Languages will change when their speakers form new groups. This adaptation follows the simple rule 'You talk like who you talk with.' As some communities stopped talking *with* other communities, they gradually stopped talking *like* them. New languages were formed. Although these changes take place homeopathically for natural reasons, their effects over time have been profound, rendering communication impossible today between folks whose ancestors spoke the same language.

We can see this principle at work in the fact that almost every language today has dialects – forms of the language where different pronunciations, grammatical constructions, meanings, words, and so on mark the boundaries of distinct speech communities. The most common type of boundary marked by these linguistic features is topological.

Look at a map of Europe. Whenever you see rivers or mountain ranges, there is a strong probability that the languages on either side of the boundaries are distinct, even though at one time they were probably the same. Spanish and Portuguese are divided by rivers and mountains crossing the Iberian peninsula. French and Spanish are separated by the

Pyrenees mountains. The mountains of northern Italy divide a multitude of small dialects and languages from one another – languages like Piedmontese, Trentino, and Bielese. If a river, a forest, or a mountain range keeps you from talking to people, soon you will not talk like them. When enough time has elapsed without talking to them, your languages will no longer be mutually intelligible – you will not be able to talk to one another.

So the history of languages in Europe is a family history – a history of the Indo-European family. And many cultural and linguistic similarities remain between European languages because of their regular contact and their descent from a common source. These similarities have had an interesting effect on the history of western philosophy. In this context it is not surprising to learn that much European philosophizing and research on languages, from the efforts of the Port-Royal Grammarians of seventeenth-century France to the extremely influential Prague Circle linguists of the early twentieth century, was centered on the idea that languages share a common universal grammar, or at least that most languages have a large number of properties in common. From the Prague Circle, this idea spread to the USA with the great European humanist Roman Jakobson profoundly affecting the theorizing of the young 'generative' linguists of MIT, led by Noam Chomsky, the founder of modern linguistic nativism.

The linguistic diversity of Europe, while much less pronounced, perhaps, than in other parts of the world, is the result of the same factors that produce difference between languages elsewhere. Topography is only one of the boundaries that can result in new communities and usher new languages into being. There are many others, including political boundaries, religious boundaries, age boundaries, gender boundaries, ethnic boundaries, and economic boundaries.

Some of these boundaries can work together. For example, Hindi – a major language of India – and Urdu – a major language of Pakistan – are more or less the same language from a purely linguistic perspective. They share more than 85 per cent of their vocabulary, nearly identical grammars, and so on. If you speak Urdu, you can understand Hindi and vice versa. Both languages descend from the same language, Indo-European, that most European languages, Farsi (Iran), and many other languages come from. But for religious reasons Urdu, which is spoken predominantly by Muslims, is written in Arabic script while Hindi is written in Devanagari script. For religious reasons also Urdu has borrowed many words and expressions from the Arabic Koran. The religious divide ultimately forced a political separation of the original India in 1947, resulting in two nations, the Dominion of Pakistan and the Union of India. This politico-religious division (tracing back to the Mongol invasions of the subcontinent) has divided the communities of speakers of the one language, Hindi-Urdu, such that in another century it almost certainly would issue in distinct, mutually unintelligible languages. The process is slowed now, however, and could be blocked by new technological forces – television, movies, and the internet – that could keep each community hearing and understanding the language of the other for the foreseeable future.

Meanwhile, a smaller religious wedge is being driven between communities of the Xavante people of central Brazil. Xavante communities currently divide into those that read the Xavante language in the writing system devised by Catholic missionaries and those who read Xavante in the Protestant orthography. The Xavante villages are often divided between Catholic, Protestants, and traditionalists. The division was effected by the attitudes of the missionaries who introduced them.

In cases like this, unfortunately, a common consequence is that each community is likely to see the other community as inferior, rebellious, or otherwise deviant, such as failing to preserve the original linguistic heritage in some way. There are those, for example, who look askance at American English as somehow less 'pure' than British English. Or Mexican Spanish as inferior or 'corrupted' when compared to Iberian Spanish varieties. For example, when Americans hear the variety of British English spoken by the Royal Family, so-called 'Received Pronunciation' or 'RP', they often categorize the British dialect as superior to their own. Some Americans might say that Prince William's speech shows how English *should* be spoken.

But when some American English speakers listen instead to, say, poor Southerners, or teenagers, or hip-hop artists, they will often judge these speech varieties as inferior to, or at least somewhat out of kilter with, the way American English should be spoken. But judgments of the inferiority of other social groups have no scientific basis. They are, rather, the reflex of bigoted judgments about the *speakers* of these varieties. If adults think that teenagers are inferior, they will think of their dialect as inferior. If a listener thinks African-Americans are inferior, they will judge them and any distinctions perceived in their dialect as inferior. And so on. But these are intellectually unjustifiable. There is absolutely nothing in the dialect spoken by the artist 50 Cent that is linguistically inferior to the dialect spoken by Queen Elizabeth.

Recognizing this, in 1996, some educators in the Oakland, California, school district decided that they would no longer submit to unscientific and biased attitudes towards the speech patterns of African-American school children. Previously, the speech of African-Americans had been labeled 'Black English' or 'African American Vernacular English.' But the Oakland educators decided to use an alternative term, coined by participants at a conference in St Louis in 1973. This term was a clever combination of the words 'phonics' and 'ebony', to produce a new label, 'Ebonics' or 'black sounds.'

These folks recommended that special programs be implemented to provide the same degree of public assistance to Ebonics speakers as to speakers of other foreign languages. The mistake that the movement made was to overplay their hand and to call Ebonics a distinct *language* rather than a *dialect* of English. The evidence is clear that Ebonics is but one of many English dialects. Therefore, although the movement was justified in fighting against prejudice and arguing for a more comprehensive and helpful educational policy, it hurt its own cause by calling Ebonics a language. On the whole, however, Ebonics was an appropriate and timely effort to help teachers, students, the greater Oakland community, and the nation at large to recognize that the language spoken by African-Americans is a tool and a delimiter of group identity, just like any other dialect or language. Everyone needed, and still needs, to know that Ebonics is not inferior to any other dialect of English. It is no more and no less 'correct', nor any more nor less 'complex,' than any other dialect of English, even the Queen's.

Unfortunately, the effort backfired. The entire gallant attempt was ridiculed and caricatured as just one more attempt of bleeding-heart liberals to tell a bunch of slackers that it was OK to use 'bad grammar.' Part of this backlash was due to overstatement by supporters of Ebonics.

Another reason for the backlash came from some educators' wish to use Ebonics in the classroom as well as 'standard' Californian English, in order to help students master both dialects of English. Still, this backlash was based on more egregious and unscientific errors than the Ebonics movement. Nevertheless the episode does show how helpful it would be for societies if people recognized that language is a human-invented tool to talk with, not an inspired medium that only degenerates over time as humans mark it with their uncouth tongues. It is in fact revived by change in each new community and remade afresh for every new group purpose.

There are many books and scientific papers demonstrating my points above about the relative parity in complexity and expressiveness of Ebonics with any other dialect of English. That is why it interests me that people are so ready to pass judgments on linguistic matters they know so little about. This is a widespread phenomenon, of course, on a wide range of issues, from what constitutes 'correct speech' to the controversy over global warming. But language seems to come out particularly badly in this smorgasbord of uninformed opinions. Everyone seems to have a view about language: who speaks it correctly, which languages are superior, and so on.

One reason it is so tempting to make pronouncements about 'proper English' and the like is because our current educational system does not help students understand why different English dialects exist and what they are like. The 'one true way' philosophy leads to anti-social and ludicrous conclusions, such as the backlash against Ebonics, a reaction which was largely ignorant, anti-social, and anti-pluralistic.

Other language issues in the US educational system are similar. For example, I was required to study Spanish from sixth grade through tenth. This was welcome and useful because more than 50 per cent of the students in my school were either Mexican-American or Mexican. The school was located eight miles from the border between Mexico and California. From my earliest memories, I liked Mexican people, Mexican culture, Mexican music, and, especially, Mexican food. So learning Spanish was like receiving the key to all these things I loved. Suddenly I was able to talk to my friends as they talked to each other.

Many of my Mexican friends, on the other hand, were not able to say the same thing about Californian culture – they were not able to talk to us as we talked to ourselves. The reason was that they were either very recent immigrants and native speakers of Spanish or native-born American citizens of Spanish-speaking parents who had not spoken English until they reached school. These children were thrown into an all-English academic environment, aside from Spanish class. Some tried very hard. Others felt so alienated that they withdrew in spirit from school. Some were successful students. But many were left trailing native English speakers in achievement.

The state of California launched a bilingual education program to help native speakers of Spanish. This was a sensible move because Hispanics make up nearly 40 per cent of the population of the state and so California's long-term economic health depends on Hispanics and Anglos, among others, working side by side. But as in Babel, the citizens of California cannot work well together if they do not speak a common language. And students from one language who only have access to education in another, with no special help, are going to be disadvantaged. And if they are disadvantaged, the entire state is disadvantaged. This is why opposition to bilingual education is so short-sighted.

Yet even when programs like Ebonics and bilingual education do get up and running, they suffer from the 'one true way of speaking' prejudice. The one true way idea lingers perhaps because of books that have come to take on a biblical-like role in the educational system. One such is *The Elements of Style* by William Strunk and E. B. White, originally published in 1918 and which, for more than fifty years, was used by a large percentage of English teachers in America as the last word in English grammar. To be fair, the book never claimed to be scientific. Its purpose was to standardize American English usage in writing and, to a lesser degree, speaking. Unfortunately, its prescriptive pronouncements are based on little more than the authors' opinions, uninformed by science.

That is why it was that in the April 17, 2009 issue of the august *Chronicle of Higher Education*, the eminent linguist Geoffrey K. Pullum published an entertaining, but dead-on accurate scientific demolition of Strunk and White's tome. The title of his piece was '50 Years of Stupid Grammar Advice.' And his conclusion? That 'both authors were grammatical incompetents.' His review should destroy not only the authority of *The Elements of Style,* but also the entire notion that one language or way of speaking is in any scientific way inferior to another.

Yet humans have always wanted to believe that their way of speaking is better than others'. The Pirahãs, for example, call their language **xapaitíiso**, 'straight head,' and refer to other languages as **xapaigáiso**, 'crooked head.' And we still hear, from time to time, statements to the effect that French is a more 'romantic' language than German or that German is a more 'scientific' or 'cold' language than English, or indeed that some languages are more logical than others. Culture affects our views of languages as well their forms. But these views have no scientific basis. What we can say is that each language fits its environment by enabling its speakers to speak in a way that its culture values.

But the fact that all languages are equal does not mean that they are all equally complex or equally versatile. What I mean here by 'equal' is the fit of each language to its cultural niche, as a result of standard human intelligence and evolution. Languages fit their cultural niches and take on the properties required of them in their environments. That is one reason that languages change over time – they evolve to fit new cultural circumstances.

It seems that the reason so many western scientists think that all languages are equally complex or versatile is due to history and a confusion about the meaning of 'versatility.' All languages change. They do this precisely because they are tools and are under constant pressure to fit the conditions of their use. As conditions change, languages change. But this does not mean that *languages* are versatile. It means that *humans* are versatile in their usage of language. On the other hand, if we were to take a snapshot of the grammar, vocabulary, and other features of several distinct languages at a specific point in time and compare them, we would discover that some languages are better than others at expressing some things, and that some are capable of expressing things that other simply cannot. For instance, as there are no numbers in Pirahã, that places obvious limits on computation in that language. On the other hand, if the Pirahãs one day were to face changed circumstances such that their culture now needed numbers, I predict that they would have them. And the reason for this is because languages do not create numbers and the like, people do. All the evidence available to us in centuries of cross-cultural research is that human beings are equally versatile and can, through learning other languages or changing their own, be equally expressive. Languages change to become as versatile and expressive as their speakers need them to be. But the source of the versatility is the language's speakers – their minds and their cultures.

In America, the history of language studies took a different direction than studies in Europe. In America, because of the hundreds of indigenous, non-Indo-European languages of the American Indians, the focus was initially on the differences between languages, trying to describe each language in its own terms. It was in America, not Europe, that the field of descriptive linguistics – the branch that produces grammars, dictionaries, and ethnolinguistic studies – was born, although European missionaries had long engaged in writing grammars, compiling dictionaries, and otherwise describing individual languages.*

In American descriptive linguistics, the practice of field research to understand language differences and to preserve records of dying languages was paramount. At the time that Columbus led the European invasion of the Americas, there were more than three hundred languages spoken in North America and these languages had been separated from the languages of Asia and Europe for tens of thousands of years. Following the likely African monogenesis, all of the subsequent influences and histories of these languages had therefore come about independently of European languages. As a result, American Indian languages had come to differ dramatically in sound, structure, and range of meanings they expressed from the languages of the Europeans.

Although American research started out inferior to the older research traditions of Europe, the linguistic diversity of the Americas gave American linguists a healthier appreciation of the differences between languages and of the need to understand these than their European counterparts. At least until the 1950s, in fact, it would not have

---

* Although Indian, Muslim, and Chinese traditions for the study of language also existed, these traditions do not fit neatly into the western categories.

been an exaggeration to say that, while Europeans were trying to understand the similarities of the world's languages even as they conducted field research in Africa, Asia and South America, North American linguists were trying to understand the implications of language diversity. A universal grammar would not have occurred to most descriptive linguists. Perhaps it was less interesting to comment on similarities between Cherokee, Iroquois and Diegueño, for example, than to understand the novelties they brought to our understanding of the limits of human language.

From their vantage point, American linguists believed that studying American Indian languages should go hand in hand with the study of these cultures, since they noticed that these cultures and languages exerted a mutual influence. American linguistics was born from American anthropology. But many Europeans looked at commonalities among languages and attributed these primarily to a psychological homogeneity among humans. Linguistics in Europe was thus often linked to psychology, not anthropology.

This multi-stranded history is why in American universities today we find linguists in anthropology departments, in independent linguistic departments, or tightly connected to cognitive science or psychology departments. Looking at the differences between languages, though, can give us a sense of the invisible hand of culture shaping them as tools, just as it shapes all our tools.

If language is a tool shaped by societies to bring coherence to community lives, perhaps there are other tools which have a similar social purpose, tools that might provide some insight into the development of language. One such tool, at one time common to all humans, is grooming. Humans and other primates engage in social grooming, an activity in which individuals in a group clean or maintain each other's body or appearance. Though this tool has been abandoned in many modern cultures, at least in its original form, other tools have taken its place.

Robin Dunbar discusses the importance of grooming in his 1998 book *Grooming, Gossip, and the Evolution of Language*, in which he promulgates the view that language is a less expensive form of social grooming, in terms of group time and resources, than the actual physical grooming that forms such a large part of other primate social behavior.

Human communities are larger than their counterparts among other primates, so physical grooming is too time-consuming and otherwise costly to maintain; language provides the contact and reinforcement of social ties that would otherwise be provided by physical grooming. Now, I don't think for a minute that language originated from social grooming. But this hypothesis does make some sense – one function of language is similar to grooming, in that it brings people together and strengthens feelings of group identity.

In light of the grooming hypothesis, it is worth bearing in mind two arresting facts about the language and culture of the Pirahã people. The Pirahãs still engage in group grooming. They live in small communities in which social grooming is easy and natural, reflecting the behavior of the smaller bands that early humans might have lived in. Interestingly, the Pirahãs also lack so-called phatic communication, or small talk – they have no words for 'hello,' 'goodbye,' 'how are you?' and so on. Phatic communication is even called 'linguistic grooming' by some linguists and anthropologists, who believe that it is related in function to physical grooming. Phatic communication carries little if any real information. It primarily demonstrates recognition of another member of one's society. That is why when you ask someone, 'How are you?' you are not really expecting them to say anything other than 'Fine. And you?' We ask these questions merely to recognize and 'stroke' one another linguistically. Thus it is not dishonest to ask a question expecting a formulaic answer. Phatic questions do not look for information. They communicate social belonging.

The Pirahãs groom each other during the day in their huts, usually just a couple of members of a single household, and in the late afternoons in somewhat larger groups. In their huts one person – usually a member of the immediate family – places their head in another's lap, and the other person looks for lice, stroking their hair, and frequently touching them softly in the process. My children used to enjoy participating in these afternoon grooming sessions and I myself allowed younger Pirahã children to go through my hair for lice. It was very relaxing, though the Pirahãs were disappointed that I didn't have lice, so they would 'plant' them on my scalp. In addition to pulling them out of each other's hair, they like to eat them. Grooming can be nutritious.

Thus they cultivate and harvest them. One Pirahã man told me that he 'really eats them,' which is the equivalent of saying 'they're yummy.'

In the evenings it is common for groups of anything from three to ten or even more people to sit single-file, back to front, on the ground or on a log, and groom one another. Most often, while grooming they talk in musical speech, a special mode of speech for communicating new information. This activity creates a relaxed atmosphere of mutual touch, conversation, and cohesion. The natural coupling of grooming and laughter, singing, speech, and moderate playfulness as a focal point of everyday village life is an effective tool for community-building.

It could be tempting to rush to the conclusion that the Pirahãs lack phatic communication precisely because they still practice social grooming. But it is not at all clear that that would be correct. Aside from the fact that there are other groups that have both social groom-ing and special forms for phatic communication, such as among some groups in the Philippines, the Pirahãs themselves co-opt normal speech expressions for phatic communication frequently.

For example, in the morning when men leave to hunt or fish, women will yell out to them to be careful, to kill lots of fish, to come back by dark, to bring them the large monkey they saw three days ago when they were gathering Brazil nuts, and so on and so forth. None of these things that are yelled out to the departing men is actually intended to elicit a response. This yelling is just a way for the women to say, 'Goodbye. Have a nice day,' using everyday words for phatic connection. In the same way, when Pirahãs leave the group gathering in the evenings to go to sleep, they often say things to each other like, 'Don't sleep, there are snakes' or, 'Don't sleep, there are jaguars.' Although there are certainly jaguars and snakes around the village, the primary function of these leave-taking pronouncements is to communicate the equivalent of American evening leave-taking expressions such as 'Good night. Sleep tight. Don't let the bed bugs bite.'

Still, the Pirahãs do not engage in phatic communication as often as some westerners do. They almost never say anything corresponding to 'Thank you' or 'You're welcome,' for example. And if one individual gets up to leave a group, whether he is going to the field or to bed or off on a two-day fishing trip, most people will not say anything to him, not even members of his family. We have to conclude that the Pirahãs do communicate phatically, but that they do this much less frequently than other cultures.

None of this is random or coincidental. Human cultures evolve and develop tools as they need them to meet their cognitive, emotional, biological and social needs. Group cohesion is reinforced by special phatic forms and phatic communication more generally. One idea about the relationship between communication and grooming is that as societies become larger, social grooming will become less common and 'linguistic grooming' will take its place. More complex societies do tend to use language differently than simpler societies.

By 'simpler' and 'complex' here I mean something quite specific. A better way of getting at what I mean would be to begin with the ideas of 'societies of strangers' *v.* 'societies of intimates.' (I borrow these phrases from the writings of Tom Givón.) In societies of intimates, like the Pirahãs, everyone knows everyone, there are no strangers. Societies like the United States, on the other hand, are mainly composed of strangers. We all meet strangers every time we leave our house. Societies of intimates tend to be smaller, more isolated, and less aided by technology. Societies of strangers are larger, more widely distributed geographically, and often more united by technology than by direct communication.

It makes sense, therefore, that special linguistic tools could be used in place of other tools, such as grooming, depending on the society. If so, this would illustrate once again that language and other forms of cultural behavior can complement, shape, or supplement one another.

Commenting on my work in the January 2010 issue of the German publication *GEO Magazine*, Noam Chomsky was asked about my suggestion that culture affects grammar. His answer is revealing: 'You begin to speak as a child. All language processes are acquired culturally? But what does a three-year-old know of culture? Nothing!'

On the one hand, I am not aware of any claim that all language processes are acquired culturally. Since there is very little research on the effects of culture on language acquisition, no one is in a position to make such a statement. On the other hand, this remark shows a profound lack of reflection on the nature and acquisition of culture, which surely begins even before language acquisition. Here are some examples of the cultural values we learn early on, independently of language:

*Hygiene*: Children learn whether it is OK to eat with the dog, off the ground, off the table, off of plates, to wipe their bottoms after defecation, to wash their hands occasionally, and so on.

*Food*: They learn what is normal to eat, what is not eaten, how to eat hot foods and to pick them out of a fire, what kinds of seasoning to use, whether to eat lots or a little of sweets, to drink water from the river or from a bottle, and on and on.

*Manners*: Children also learn whether or not to acknowledge others, how to act when an adult speaks to them, how or whether to indicate their desire for more food, and so on.

*Emotional security*: Long before they acquire language, young children learn whether to draw their emotional security exclusively from their mother or their mother and father or from all members of the community or from some select group within the community. Children observe their parents' faces and actions in difficult situations and learn to act and react as their parents do. This can vary tremendously from culture to culture, as researchers under the direction of Dr Heidi Keller of the University of Osnabrück, Germany, have shown in numerous studies.

Still other values acquired by children without language include all kinds of appropriate behavior: when to sit still; when to be quiet; how to express affection; what kinds of clothes to wear, if any; the length and intonations of verbal interactions; how to hold and manipulate your body (resulting in what might be called greater or lesser gracefulness); which tools are used and for what; interpersonal etiquette (for the Pirahãs, throwing snot off your hands on to others when you have a cold is not a bad thing, so long as it is not done intentionally); attitudes towards death, trustworthiness, barter and exchange, relatives (kinship system); two-dimensional *v.* three-dimensional objects; and so on.

Having established that it is reasonable to assume that culture acquisition begins early, let us turn back to language. There are many cognitive and cultural tasks for which language is vital. One of the most important of these is categorization.

Differentiation of the things around us is one of the first pillars of language. We need to sort the world into different categories to fix our points of intellectual and social relevance.

Philosophers wonder whether the things we categorize are there because we have words for them or because there really is an underlying distinction or perhaps they are just imagined. That is, do languages agree about what the world consists of because the world consists of those things or because humans share limitations that lead them to perceive the world in similar ways?

People can perceive and talk about the world in radically different ways. The Pirahã expression for 'Don't move!' means literally 'Not into the jungle! Stay!' Why would you tell a little child sitting in a canoe on the middle of the river not to 'move into the jungle'? What might the Pirahãs mean when they say 'jungle' in these situations?

More examples might help approximate an answer. One use of 'jungle' that helped me better understand this word occurred when the Pirahãs watched on my computer with me a National Geographic DVD about living in the snow. They commented that 'those people live in another jungle.' 'What jungle?' I asked myself. 'It's just barren tundra, for heaven's sake.' When a Pirahã heads to the forest to hunt, he or she will say, 'I am going to the jungle.' The word for all of these 'jungles' is the same: **xooi**. Clearly what I was initially calling 'jungle' does not quite mean what I thought. I translated this using my own categories, rather than Pirahã categories. The best translation that I have been able to come up with is 'environment,' if what we want to do is find an equivalent that is general enough to include all the uses. 'Don't move in the environment' means 'stay still.' To the Pirahã the tundra is 'another environment' but jungle is their own basic environment, hence their use of the word in a more general sense.

Language is crucial for categorizing the world around us. It is used as a cultural tool in this sense. For example, although I am not the most organized or smartest person around, I can still sort things better than my dog. Although my dog can and does sort some things. He can pick out what is edible from what is not edible, sorting food and non-food into separate piles. He can separate meat products from other foods, thus sorting the edible world into meat *v.* non-meat. And he can pick me out from all other people, showing his ability to sort Dan *v.*

non-Dans, and so on. There is no question that he can sort. Just as all animals can.

Greyfriars Bobby of nineteenth-century Edinburgh fame is another, better-known case of a dog-sorter. According to legend, he daily picked out his dead owner's grave from all others. Of course, if Bobby had been a person he would have been committed to a mental institution for continuing to wait for a dead owner that was never coming back. People knew the owner was dead – just not the cute little terrier. The lesson here is that Greyfriars Bobby had at least some cognitive resources reminiscent of sorting, a prelinguistic precursor to categorization. But compared to you or me, dogs are not good at categorizing. They can do it, but they lack language, a major cognitive aid to sorting and categorizing. Indeed, dogs are not very good at any number of other things that we take for granted.

There are many reasons besides language for the disparity between dogs and humans in cognitive tasks. First, we and dogs have different kinds of brains which evolved to do different jobs. My dog's brain is much better than mine at sensing strangers, finding its way back home, and so on. Saying that humans are better at some other things than dogs is little more than a restatement of what we already know: humans and dogs evolved to fit different niches; they have different ways of surviving. And language is part of our niche-fitting. Language in particular works well for categorizing and classifying the components of the world around us.

We have evolved to create and store concepts through signs and to recognize relationships between the signs so formed. When we use, learn, or invent a noun, for example, we have necessarily made a generalization. The noun 'dog' can refer to a specific dog, as in 'that dog,' but more importantly it relates all dogs to one another as members of a class, as individual exemplars of a general concept, dog. No culture in the world, no non-pathological human being anywhere, will be found without the ability to generalize. I have made the case in numerous publications that in the Pirahã culture certain kinds of linguistic generalizations are avoided, such as numbers and color words. But if any culture lacked entirely the ability to generalize, it would be incapable of learning human language. This is why *no* culture can override the core, minimal demands of communication or life in the world. If it did then that culture would itself cease to exist. Just as every noun is a generalization about things, so every verb is a generalization about events.

But each culture determines which generalizations are the most important to it, its vocabulary reflecting its priorities of knowledge. Lexical distinctions – types of nouns, verbs, modifiers and so on – are established in order to communicate about topics valued by a particular culture. One of my favorite examples of this cultural dictionary management comes from kinship systems –that is, schemes for classifying and recognizing relatives.

The first interesting reference to relatives in western literature is in the Bible. Genesis 4:16 and 17 tells us that one of Adam and Eve's sons, Cain, went to the land of Nod and took a wife. Until this point Genesis had led us to believe that Cain and Abel were the first children ever born, the *only* offspring of Adam and Eve. But if Adam and Eve were the first people then the people of Nod must have also been their offspring. So Cain had to marry a relative, according to this story. From the moment humans began to mate and procreate, they began to think about whom they should mate and procreate with. Humans classify each other with respect to their marriage relationships – potential marriage partners, impossible marriage partners, children of partners, relatives of partners, and so on. Eventually 'partner patterns' emerge in every culture. And in modern times the study of these patterns has become a principal domain of study in anthropology.

In 1871 Lewis Henry Morgan published his seminal, pioneering work, *Systems of Consanguinity and Affinity of the Human Family*. Perhaps Morgan's most important contribution in this book was to demonstrate convincingly that kinship was not merely a way of labeling pre-existing biological relationships but a classification system for integrating biological and non-biological relationships for different yet overlapping purposes determined by individual cultures. Kinship always serves to identify and establish relationships that are valued culturally. Morgan developed a theory of kinship based on descriptions of the basic organizing principles found in six representative types of kinship systems: the Iroquois, the Crow, the Omaha, the Dravidian, the Eskimo, the Hawaiian, and the Sudanese.

Morgan used these names not only to describe those cultures' patterns but also to classify all the world's kinship systems. American kinship, for instance, is an example of the 'Eskimo' system because the American system is nearly identical to that system. Hawaiian was thought to be the simplest system known and Sudanese the most complicated. Because all cultures have kinship classifications and because of kinship's important role throughout a society, generations of anthropologists since Morgan have discovered that kinship relations can teach us a great deal about the relative roles of human nature and cultural variation and this variation's cognitive effects. So let us examine some of these kinship systems more closely and add a different system, the Pirahãs', to the mix, to understand what kinship might have to tell us about the range of cultural and linguistic tools available to humans to categorize their relationships.

In the Iroquois system, different genders and generations are distinguished (father, mother, sister, brother, grandfather, grandson). Initially the Iroquois terms might not seem all that different from English ones. But look more carefully and you will discover a significant contrast between Iroquoian and English kinship. In Iroquois kinship, parental siblings of the same sex are considered blood relatives, while parental siblings of the opposite sex are not. For example, you would refer to your father's brother as 'father' and your mother's sister as 'mother'. Your mother's brother, however, would be called 'uncle' *not* 'father' because a mother and her brother do not share the same gender. Likewise your father's sister would be referred to as your 'aunt', not your 'mother'. The children of same-sex *v.* different-sex siblings of parents are also named differently. So your father's brother's children are your 'brothers' and 'sisters' but your father's sister's children are 'cousins'. More technically, your father's sister's children and mother's brother's children are 'cross-cousins' while your father's brother's children and your mother's sister's children are 'parallel cousins'.

Kinship terms do more than merely categorize relationships like cross *v.* parallel cousins, however. They also play a regulatory role in societies. We can see this in Iroquois kinship, where the children of *ego*'s father's brothers and mother's sisters (*ego* is the conventional reference point used in descriptions of kinship – it refers to one's self) are *not* to be married, while the children of *ego*'s father's sisters and mother's brothers are ideal marriage partners.

In the American system, which is an example of the Eskimo kinship type, there is no distinction between parallel *v.* cross-cousins. None are desirable as mates in this system. In earlier generations this taboo against marrying cousins was not so strong. Charles Darwin, for example, married his first cousin, Emma.

To bring some clarity to who we are allowed to marry, American and English kinship used to make fine distinctions between types of cousins, though these distinctions and the marriage taboos they represented now seem to be disappearing. Although English does not distinguish parallel and cross-cousins, it does distinguish 'first', 'second', and 'third' cousins and so on – because it recognizes what is called 'degree of relationship.' English further distinguishes cousins by the concept of 'removal', cousins once removed, twice removed, and so on.

These finer distinctions are worth understanding because they tell us a good deal about how language and culture work together. Before discussing these, though, it would be useful to clarify what a cousin is. Cousin is a catch-all term in English which refers to anyone with whom one shares a common ancestor. When there is a more specific term, though, we use that and not the term 'cousin'. For example, because there is a specific term for my female parent, 'mother', I don't call her my cousin, even though we share a common ancestor. For the same reason, I do not call my male parent 'cousin', because 'father' is a more specific term for him. To repeat, 'cousin' is used when there is no more specific term.

With that background, we can explain how both British and American English express 'degrees of relationships.' First cousins are any individuals who share grandparents. Second cousins share great-grandparents. Third cousins share great-great-grandparents, and so on. In the American system, at least, the use of 'second' and 'third' cousins is becoming less and less common, to the point that the average American probably uses the terms unsystematically – as terms they have heard, not concepts they have mastered. These lexical tools are losing their utility in modern American society. This reflects changing American culture values and their effect on the language.

In addition to the degree of relationship English also refers to the degree of 'removal' of cousins. This concept is about the generational distance between cousins. My first cousin's child is my first cousin 'once removed.' My second cousin's child is my second cousin 'once removed.' My first cousin's grandchild is my first cousin 'twice removed.' And so on. Any disuse and lack of familiarity with these terms in the American kinship system is a result of their reduced usefulness to our present-day society and culture.

'First cousin' continues to be an important term, however, because of the widespread belief, not entirely unfounded, that marrying relatives as close as first cousins will in some way produce inferior offspring. This is indeed a possible outcome. It is not just that marrying close relatives makes the gene pool shallower; the potential effect is even more negative than that. Inbreeding – sexual relationships with close relatives – can cause a higher than normal degree of congenital birth defects. It does this by increasing the likelihood of propagating unhealthy or undesirable recessive genes in a particular family. Interestingly, every known society has tacit or explicit rules against mating with relatives that are too close. These rules have become known as the 'universal incest taboo' (the word 'incest' comes from a Latin word meaning unchaste or sexually impure).

Kinship systems are therefore cultural tools that regulate marriage relationships and order other aspects of interpersonal relationships in societies. As the members of a culture internalize such tools, the tools come to affect that culture's members cognitively, literally shaping their thinking about some things. That is why if you use an Iroquois kinship system, you are going to talk and think differently about someone you call brother or sister than someone you call 'cousin.' You will think of the latter as a potential spouse but of the former as someone under the incest taboo.

Kinship terminology shows us that the various influences of language, culture, and cognition on our mental and social lives are hard to tease apart. Language, mind, and culture often work together to produce unseen yet strongly felt limitations on our thinking, behavior, and inter-personal relationships.

For decades, anthropologists believed that the Hawaiian kinship was

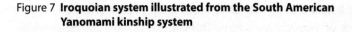

Figure 7 **Iroquoian system illustrated from the South American
Yanomami kinship system**

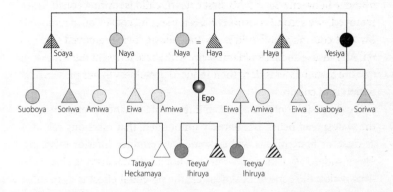

the simplest system. In this pattern, there is a very small set of terms. There is a word for 'father', which refers to any male of *ego*'s father's generation; 'mother', any female of *ego*'s mother's generation; 'brother', referring to any male of *ego*'s generation; and 'sister', meaning any female of *ego*'s generation. Then there are terms for 'son' and 'daughter', applied to any people of *ego*'s children's generation. The Iroquois and Hawaiian systems are illustrated in Figures 7 and 8.* The Iroquoian system is illustrated by the kinship terms from Yanomami, a South American language of the Orinoco river system which straddles the border of Brazil and Venezuela.

Let me briefly explain how to read a kinship chart. Individuals of the same generation are on the same horizontal level. The equals sign, =, connects two individuals that are married. Women are represented by circles and men by triangles. Vertical lines indicate children. A vertical line descending from an equal sign indicates a child of that marriage (so in the first chart below, *ego* is the child of 'naya' and 'haya'). *Ego* will call both his or her father and their father's brother 'haya', for example, in the Iroquoian system. And she or he will call their mother and their mother's sister 'naya'. *Ego*'s brother and sisters are connected to the same equal sign. *Ego*'s cousins are those at the same horizontal

---

* These were all created by Brian Schwimmer of the University of Manitoba.

## Figure 8  **Hawaiian kinship system**

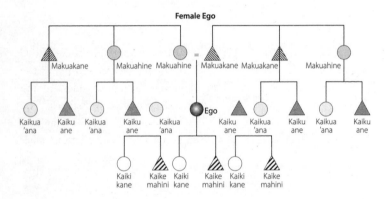

level, but not connected to *ego* via the same parents (those connected by the equals sign).

The traditional Hawaiian culture has been described by some as freer in marriage and intersexual relationships than other cultures, owing, it is assumed, to its less restrictive kinship system. In the chart here, we see that male relatives of the parent generation are distinguished only by gender. So *ego*'s mother is *Makuahine* and *ego*'s father *Makuakane*. But *ego*'s aunts are also *Makuahine*, the same as for 'mother,' while *ego*'s uncles are *Makuakane*, the same as for 'father.' All terms of *ego*'s generation are the same, whether for 'brother' or 'sister,' *Kaikua*. In the generation below *ego*, all terms are also the same, *Kaiki*. Additional terms can be used, as in Pirahã.

## Figure 9 **Pirahā kinship terms**

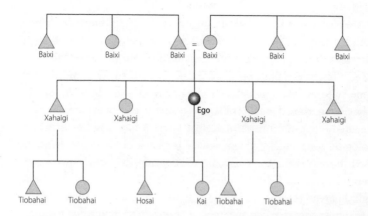

But it turns out that the Hawaiian system is not the least restrictive system. Pirahā kinship is even less so. The Pirahā system is similar to the Hawaiian system, except that it does not even distinguish gender for the generation above *ego*. Pirahā kinship is diagrammed in Figure 9. The terms are **baíxi** 'parent,' **xahaigí** 'sibling,' **hoísai** 'son,' and **kai** 'daughter.' Occasionally Pirahās use non-kinship terms for kinship relations, such as when *ego* calls his father **ti xogií** 'my big' or **ti xibígai** 'my thick.' These expressions, like **baíxi**, can be used to refer to parents or grandparents or, in some circumstances, to anyone of any generation who has power over *ego* in some way.

This, the most unrestrictive kinship system known, seems to be correlated with Pirahā sexual and marital practices. For example, any Pirahā male can have sex with or marry any female except for a full biological sister (half-sisters seem to be OK), mother, grandmother, or daughter. Pirahās marry step-parents, cousins, half-siblings, and others that the average American would not. Like most other discoveries made in field research among little-studied groups like the Pirahās, there are multiple lessons to be drawn from each new set of facts like these.

For example, not only does the correlation between kinship and marriage show that kinship language is a cultural-cognitive tool, the Pirahā case raises a problem for any simple attempt to say that our language always constrains our thinking, a hypothesis that some call linguistic relativity.

The problem raised is that, although Pirahã does have an incest taboo – no marrying of biological parents, grandparents, or full siblings is allowed – the language lacks specific kinship terms to outlaw this. This taboo therefore cannot be linguistically motivated. This means that some cultural values as well as some universal constraints on human behavior are rooted outside of language. In the case of Pirahã, the only potential linguistic constraint against incest is found in the terms for biological sons and daughters, which are listed in Figure 9. There *are* linguistic terms for these specific relationships. On the other hand, there are no terms for 'full biological sibling' (brother or sister) or 'biological parent' (mother or father) in Pirahã, yet still no member of the community would marry anyone that bore such a relationship to them, even in the absence of a linguistic term for that relationship. So the prohibition against marrying one's sister, brother, father, grandfather, or grandmother is a cultural value with no known linguistic expression.

There are other cultural tools whose effects are more immediate and more obvious to the casual observer, however, than kinship terms, phatic language, or other linguistic tools. There is no reason I can see not to classify cultural technologies like language, theories of language, scientific methods, and so forth as tools on a par with other cognitive-cultural implements. In the corner of the room where I am writing stands my Martin acoustic guitar, with a pick interwoven within the first three strings. When I get tense or want to relax, I slide my chair back, put my guitar in my lap, and play and sing whatever comes to mind. My guitar is a tool. It helps me make music. It helps my voice stay on the melody. It helps me to relax. And so I have a guitar and I have learned to play it, in order to avail myself of its utility. Guitars or golf clubs or ratchet wrenches or bows and arrows, all are part of a society's tool kit, provided by cultural heritage and individual inventiveness.

Understanding the full range of cultural tools devised by our species requires a good deal of field research. Field research on unstudied or little-studied languages and cultures is the key to understanding the varied functions, structures, and effects of the partnership of culture and language on language. It is the only way for us to expand our knowledge of what is possible in human languages. But in field research we face a problem common to all science in one form or another – the relationship between the seen and the unseen. A complete understanding of a culture or a language should include explanations of why some phenomena are observed in the culture as well as why other things are not. Why do adult Americans vote, while pre-adolescent Americans do not? The absence of voting among the latter population would teach us, if we were conducting field research on American values, that there is a value, codified as a law, that you should be of a certain age to vote in American society.

That is an easy example. In general, however, it is much harder to say what a culture does not have than to say what it has. For example, let us assume that you are hiking along a wooded path near your home and you see a hawk flying overhead. You take a photo of the hawk, then you post the photo on Facebook. Now, your friends know you're honest and that you wouldn't fake a photo, so everyone now believes that you indeed witnessed a hawk and that there are hawks near your home. Now let's say you walk another path far from your home. On this new path you see no hawk. So you take a picture of empty sky and you post it on Facebook with the caption, 'There are no hawks in the woods here.' But this time not everyone believes you. Why not? After all, you are still their trustworthy friend. So how is it that they believed you before but they are more skeptical now?

That first photo you posted was pretty convincing. It showed a hawk to support your claim that you saw a hawk. So why isn't the picture of empty sky just as convincing as the picture of the flying hawk? Because the absence of evidence is not evidence for absence – it is much harder to make the case that something is absent than that it is present.

Or imagine that your spouse accuses you of having an affair. A picture of you being intimate with another person would be pretty convincing evidence that you are in fact being unfaithful. But a picture of you alone in your hotel room would not convince anyone that you have never engaged in hanky-panky.

In spite of such obvious difficulties in convincing others about things cultures do not have, it is possible to find evidence of absence rather than merely absence of evidence. And sometimes the absence of something teaches us *more* about the world than its presence. Sometimes it is the conjunction of the things we can talk about and the things we cannot talk about that reveals more about us then either of these alone. And the things we talk about and don't talk about can affect the way we think. Different languages and different cultures can, therefore, produce different thoughts.

# PART FOUR:
## Variations

## Chapter Ten

# LANGUAGE, CULTURE, AND THINKING

*'We dissect nature along lines laid down by our native language.
Language is not simply a reporting device for experience but a defining framework for it.'*

Benjamin Lee Whorf, *Language, Thought and Reality* (1956)

Benjamin Lee Whorf was a fire prevention engineer working for the Hartford Fire Insurance Company. But he was also a linguistics enthusiast and through his amateur fascination with language and the mentorship of the great American linguist Edward Sapir, Whorf came to the conclusion that the languages we speak affect the way we think, almost like a straitjacket affects our movements. If he is correct, our thinking patterns can only be understood *relative* to our language patterns. Hence his proposal is often referred to as 'linguistic relativity' (it is also referred to as 'linguistic determinism,' 'the Whorf hypothesis' or 'Whorfianism').

This hypothesis actually precedes Whorf. Whorf learned 'his' hypothesis from Sapir, who in turn had learned of it from his teachers in Germany. Sapir sums it up best, when he says that 'Meanings are not so much discovered in experience as imposed upon it, because of the tyrannical hold that linguistic form has upon our orientation to the world.'

The Whorf hypothesis comes in different forms – a strong form, a weak form, and a very weak form. The strong form maintains that language *controls* our thoughts. Some believe this, but few if any serious researchers give credence to this form of the Whorf hypothesis. The weak form says that language influences our thinking about the world in most tasks. The very weak form suggests that language can influence the way we think in highly specific, real-time tasks.

While some contemporary researchers today believe that any form of the linguistic relativity hypothesis is balderdash, there are many psychologists and linguists who claim that their own research supports Whorf's hypothesis.

Over a period of twenty years, I looked for things I could not find in the Pirahãs' culture and language. I looked for number words, color words, creation myths, complex sentences, quantifier words, and other things that were not turning up in my data. Then it began to dawn on me that perhaps these things really could not to be found, that they were in fact absent from Pirahã. Colleagues of mine have argued that the lack of certain categories of words – number words in particular – affects the way the Pirahãs think and therefore that Pirahã supports linguistic relativity. For example, it has been shown in three separate sets of experiments, by researchers from Columbia University, the Massachusetts Institute of Technology, Stanford University, and the University of Miami, that not only do the Pirahãs lack number words but they also do not count. Some have therefore concluded that the Pirahãs do not count because they do not have number words. This conclusion would support the Whorf hypothesis.

Some claims for linguistic relativity that have been based on Pirahã are wrong, though. For example, I believe that the Pirahãs' lack of counting and their lack of number words are both caused by a cultural taboo against unnecessary generalizations beyond the here and now. This may or may not be correct. Therefore I still lead teams of psychologists and linguists to the Amazon to test not only my own but others' hypotheses about the relationship between language and thought. More evidence is always useful for evaluating any scientific hypothesis, especially a controversial one.

I first learned about the Pirahã language through the writing of Steve Sheldon, who served as a missionary to the Pirahãs from 1967 to 1976. He claimed that the language had four color words: **biísai** 'red,' **kopaíai** 'black,' **koobiai** 'white,' and **xahoasai** 'blue' or 'green' (one word for both). When I first arrived among the Pirahãs, I used these expressions regularly and they seemed to work just as Sheldon had predicted, so I did not think about them again for several years. After all, about 80 per cent of the language still needed to be analyzed, so why invest time and intellectual energy reinvestigating aspects of the language that already seemed to be understood?

Eventually, though, I began to connect some dots that changed my perspective on Pirahã color words. My first discovery was that each 'word' for color in Pirahã was actually a phrase. For example, **biísai** did not mean simply 'red.' It was a phrase that meant 'it is like blood' – **bii** 'blood' plus **ís** 'animal/it' plus **ai** 'to be' – 'it is blood' (more loosely rendered as 'it is like blood'). And the 'word' **kopaíai** 'black' was **ko** 'eye' plus **opaí** 'unclear/opaque' plus **ai** 'to be' – 'an unclear eye.' The phrase **koobiai** 'white' breaks down into **ko** 'eye' plus **obi** 'clear' and **ain** 'to be,' or 'to be clear or transparent.' And the phrase **xahoasai**, 'blue' or 'green,' is analyzed as **xahoa** 'unripe,' **ai** 'to be,' 'it is unripe.'

Not only were these color words not words at all, but they were not the only expressions used to describe the basic four colors, black, white, blue/green, and red/yellow. To describe red in Pirahã, one can also say **xaxaí xigiábisai** 'like an urucum (red dye-producing) plant,' or **bii xigiábisai**, 'similar to blood,' or even **tomate xigiábisai**, 'like a tomato,' where Pirahãs use the Portuguese word *tomate* 'tomato' as a way to describe red (this they would do to be humorous). These phrases are all also used to describe orange, yellow and some other bright objects. **xahoasai** can describe blue things, green things, young children, unripe fruit, and so on. This is reminiscent of English where we can use the word 'green' in similar ways: 'He's too green to be sent on a mission like that' or 'This banana is too green too eat. Let's wait until it's riper.' **kopaíai** can describe a panther or other dark animal but it is not usually used for people or the color of night. Darker-skinned people are described as **biiopaíai** 'blood is opaque.' A dark night is described as **xooi** 'jungle/environment,' **tii** 'excrement,' **o** 'become,' **aa,** 'be,' **bá,** 'remain' – which literally means 'the jungle is shitty.' White or transparent objects and lighter skin, like mine, are some of the things described by the phrase **koobiai**.

Based on this closer look and reanalysis, followed by many further attempts to elicit color words over the years, I have had to conclude that Pirahã have no color words. More astounding than this lack of color words, however, is the remarkable consistency with which Pirahãs group the basic colors. It is clear after only a few minutes of research with a variety of Pirahã men, women, and children that the colors that stand out as most salient to all are the ones just discussed, what we would call in English 'red,' 'blue/green,' 'white,' and 'black.' This is interesting stuff for anyone interested in how humans categorize the world around them and how they deal with concepts that they have no words for.

So the Pirahãs are consistent in picking out the four colors just mentioned as the most salient to their lives, and they also describe these colors relatively consistently, using the expressions above, but disagree about the rest! This last fact is interesting in itself. All Pirahãs will describe any color you present them with in experiments. But there will be little consensus on the words used. For example, one person might describe gray as 'white' while another will describe it as 'like thread.' This ability to think about colors without using a specific language to do so shows that Whorf cannot be completely right. People *can* think beyond their language. Science, the ability to discover and think new thoughts and invent new terms, is an ever-present refutation of the more extreme interpretations of Whorf.

In 1969 Brent Berlin and Paul Kay, at that time both anthropologists at the University of California at Berkeley, published the influential book *Basic Color Terms: Their Universality and Evolution*. It was their claim that each language in the world could be classified as belonging to one of seven distinct stages of color development:

*Stage I: Dark-cool and light-warm*
*Stage II: Red*
*Stage III: Either green or yellow*
*Stage IV: Both green and yellow*
*Stage V: Blue*
*Stage VI: Brown*
*Stage VII: Purple, pink, orange, or gray*

They also predicted that all languages would have terms for color. In the early versions of their theory, Berlin and Kay required that all terms be simple words (that is, words not derived from other words). 'Red' is a simple word but 'blueish' is not, since it is derived from the word 'blue.'

Over the years, however, they were forced to loosen the theory to allow for the existence of systems that use phrases, rather than simple words, as basic color terms.

By their names for stages of color grouping, Berlin and Kay predict that, if a language has only four words for color, these should be red, black, white, and blue. The Pirahãs obey this cognitively – they recognize these as the most salient categories. But they have *no words* for the categories, resulting in the presence of an ability to do something (distinguish colors) without dedicated lexical tools. Again, this is unexpected by linguistic relativity as usually understood.

But notice that the Pirahã system, because it lacks words for basic colors, shows us that we can create categories for objects in the world around us without words. We saw the same point made by the presence of an incest taboo in Pirahã in spite of an absence of kinship terms to support it. If I am right about Pirahã – if they do indeed have abilities that are not reflected in their language – then it must be concluded that thought is independent of language in important ways.

The absence of number words in Pirahã tells a very different story. When I first entered the Pirahã tribe in 1977, I was told that they had a common and very simple number system, 'one,' 'two,' and 'many.' This is a common type of numerical system found in many languages around the world. Nothing remarkable. I used the words according to this analysis of the numbers that I had inherited and they seemed to work fine. Since I had gotten this analysis from a reliable source, a fluent speaker of the language with a master's degree in linguistics, I did not question it. After all, my own experience seemed to confirm it.

But then one day I thought about the word for 'baby boy' in Pirahã, **xigihíhóihi**. And I realized as I reflected more about this word that **xigihí** meant 'man' and **hóihi** was what I thought meant, 'it is one.' But it did not make much sense to think that 'baby boy' was literally 'It is one man.' So I looked for other examples of **hóihi** in my data. It turned out that the best translation for this expression, across all of the examples I had, was 'small.' So 'baby boy' meant 'small man.' That *did* make sense.

But then I thought that the **hói** that meant 'little' and the **hói** that meant 'one' must be homonyms. In a language like Pirahã, with only eight consonants, three vowels, and two tones, homonyms should be common.

Eventually, though, the data led me to the inescapable conclusion that **hói** *always* meant 'small' or 'little,' either of quantity (a small amount of manioc meal, for example) or of extent (a small man) and that there was only one **hói**, not two homonyms. This realization of mine led to further studies by psycholinguists that confirmed that the Pirahãs have no number words, nor any concept of counting in their language. This is the only community in the world, so far as we know, that has been determined to lack any numbers at all, not even the number 'one.'

People often ask me things like, 'You mean to tell me that Pirahã mothers don't know how many children they have?' Yes, that's exactly right. No Pirahã woman can tell you whether she has one, three, or six children. But while she won't know the *number* of children she has, she will most definitely know the names of all her children and she will know where they all are, about their safety, health, and so on.

It is hard for westerners to realize that math is detachable from human existence. Grade-schoolers the world over are sold the bill of goods that math is necessary to make a living. And that is correct for living in modern economies. But the Pirahãs don't have a modern money-based economy and do not otherwise need numbers.

This conclusion bothers some people. It is hard for them to accept the idea that numbers are just tools. Many nativist researchers believe that numbers are innate, that they are too basic to the human condition to be otherwise. This reluctance to view numbers in any other way strikes me as not based on research, but on an idea of what the world should be like under some researchers' assumptions. This is a misapplication of deductive reasoning – attempting to derive the particulars of the human condition from a particular theory of what humans are like.

Should it really shock us to learn that people who never need to use numbers have none? Perhaps it should. But then that very shock ought to work for us – to drive new experiments and new thinking. As with all science, we need to ask at all times whether our theories should be modified or abandoned.

The first interesting observation is that not only do the Pirahãs not count, but that experiments have further established that they do not even understand the concept of counting. Along with these experiments, further support for the conclusion that the Pirahãs are unable to use numbers or count comes from my long-term experience of trying to teach them to count by using Portuguese numbers. I discovered that, although children can learn to count up to ten fairly quickly in Portuguese, it is much more difficult for Pirahã adults.

The correlation of the absence of number words and the absence of counting contrasts, however, with absence of any correlation between words and performance as seen in the Pirahãs' behavior in the cognitive domains of kinship and color terminology. For example, we know that the Pirahãs practice incest avoidance without words to do so and that they can sort colors using the Munsell color chips made famous in work by Paul Kay and Brent Berlin for studying cross-cultural perception of colors, without using words for colors. So why, it is reasonable to ask, can they not count, even in spite of the fact that they lack numbers? *Because math is different.* Because we need to learn numbers and counting together, neither is independent of the other. This is not, however, the case for the cognition of kinship and color.

All of these important results derive from absences of words. Different absences teach different lessons. Moreover, this set of correlations between missing terms and cognitive abilities raises an even larger question, namely, what studying the Pirahã can tell us about what it means to be human.

Before we can set about answering that question, though, we need to view these findings in two larger contexts. The first context is the conjunction of Pirahã language and culture. The second context is the matrix of human cultures globally. If the historical development of each human language engages cultural values so as to help the language better serve the communicative needs and values of the community that speaks it, then each language–culture pair will teach us not only about itself but also about the bounds of human nature and experience.

I have a strong suspicion that Pirahã is not the only language in the world without numbers. My reasons include the fact that there are many cultures equally or even more isolated than the Pirahãs and that these peoples, for reasons similar to the Pirahãs', likely do not need numbers either. If we look at known languages around the world, we see many so-called 'one, two, many' systems that continued only until trade with the outside world led to a need for and introduction of more words into the language. In Australia, for example, the Warlpiri borrowed English numbers once trade relations rendered their previous numerical system inadequate for their changing needs. This is not only a clear example of culture changing the language (by the wholesale importation of English numbers) – and a consequent increase in the versatility and expressive power of the Warlpiri language – but it also raises the question of what the Warlpiri number and counting system was really like prior to contact with Europeans. If careful experiments had been done there before the language changed, then perhaps we would have learned that the Warlpiri system had been like the Pirahã system.

In my earlier work on these matters, I explained the Pirahãs' lack of number and color words as the result of an 'immediacy of experience principle,' which values talk of concrete, immediate experience over abstract, unwitnessed and hence non-immediate topics. This still seems to be the best explanation for these facts. And yet, although the absence of color and number words in Pirahã may be explained by the same cultural value, their absence demonstrates a very distinct psychological status for each. This is pretty interesting. And there is no reason we would not find similar correlations in many other languages.

However, experiments we have conducted so far are only designed to test whether all these phenomena are absent or not from Pirahã. What is harder to test is my explanation for their absence – that culture shapes grammar.

Whatever other theoretical conclusions might follow from the Pirahã number and color facts, we have learned from this community that there are languages without math and without color words. And this, incidentally, weakens the nativist claim that numbers arise ontogenetically (in each individually developing human) from the human genome.

This conclusion, that the Pirahãs show uniformity of behavior in a specific cognitive domain without a vocabulary for items in that domain, flies in the face of some contemporary research. Pirahã color categories and kinship terms show that thought can precede language in reasoning about important areas of human cognition.

But Whorf's ideas, although they clearly do have something to teach us about the connection between language and thought, seem to also attract people who want to take them much farther than either Whorf or the evidence allows.

One person has founded a movement based on linguistic relativity, though he does not use this term so far as I can tell. This is David Wynn Miller, of the Sovereign Citizen Movement, who believes that government and others have produced the current forms of our language to control our thoughts. So Miller invented his own language in 1988. He calls this new code the 'Mathematical Interface for Language or Quantum-Math-Communications and Language' or, more simply, 'Correct-Language.' Among Miller's ideas is that only nouns can have legal authority. This movement has attracted a surprising number of followers, some of who have tried to use Miller's theory in court cases. Others have even attempted to link Jared Lee Loughner, the mass killer of Tucson who shot US Representative Gabrielle Giffords among other victims, to Miller.

Here is a sample of Miller's writing in 'Correct-Language:'

~1 FOR THIS PLENIPOTENTIARY-JUDGE: David-Wynn: Miller's-KNOWLEDGE OF THESE CORRECT-SENTENCE-STRUCTURES-COMMUNICATION-PARSE-SYNTAX-LAN-GUAGE=(C.-S.-S.-C.-P.-S.-L.) IS WITH THE CLAIMS BY THE QUANTUM-LANGUAGE-PARSE-SYNTAX-NOW-TIME-WRIT-TEN-COMMUNICATION-FACTS.

~2 FOR THE EDUCATIONAL-CORRECTIONS OF THESE MODIFYING-COMMUNICATIONS ARE WITH THESE COR-RECTION-CLAIMS AGAINST THESE FICTIONAL-ADVERB-VERB-USAGE WITH AN OPERATIONAL-METHODS OF A FICTIONAL-MODIFICATION-LANGUAGE. (8500-YEARS OF THE SYNTAX-MODIFICATIONS WITH EVERY LANGUAGE)

If Miller had read Whorf carefully, he might have realized that his attempts to produce a non-oppressive language make no scientific sense. Although they follow, however tacitly, some core ideas of Whorf, they none the less miss the point that languages cannot have all cerebral influence removed from them. Although we can always profit from being sensitive to our languages, the fact that languages evolve to fit cultures and that cultures have bad in them as well as good, means that our languages will never be 'pure' in any sense.

Again, the weak form of the Whorf hypothesis claims that our languages can *affect* (not determine) the way we think. Most of the growing numbers of researchers who are investigating the effects of language on thinking have come to be known as 'neo-Whorfianists.' They believe that we *can*, when given time to reflect, think beyond the bounds of our language but that often we do not. Language affects our thinking in carrying out tasks quickly and in subtle ways.

For example, Dan Slobin, a psychologist at the University of California at Berkeley, has demonstrated that French-speakers and English-speakers perform differently at certain tasks based on the way their languages encode the concept of 'manner of action.'

In English the manner of action can be expressed directly in the verb. In French it cannot be, but must be expressed as a separate phrase:

The dog *ran* into the house.
*Le chien est entré dans la maison **en courant**.*
'The dog entered the house *by running*.'

In the English example, we know that the dog entered the house and that it did so quickly simply by the meaning of the verb 'ran.' In French we must say instead, 'the dog entered the house by running.' The consequence is that speakers of French talk about actions, translate actions, and think about actions differently from speakers of English. Speakers of English are more likely to talk about, think about, and translate manner of action as part of their linguistic and cognitive behavior than French speakers. This follows, according to Slobin's neo-Whorfian perspective, because manner of action is expressed with greater ease and frequency in English than in French and because it appears directly in the verb 'ran.' This is not a huge cognitive difference, but it is a difference and it seems clearly based on or reflected in the differences in word formation between French and English.

Slobin also showed in the 1960s that passive sentences, such as 'Bill was seen by John,' are understood by native speakers more rapidly if the meaning of the verb only allows one of the two noun phrases in the sentence to be the one doing the action. A passive sentence like 'The horse was kicked by the cow' takes a relatively long time for for native speakers to process, because both cows and horses are known to kick. So speakers, in spite of the clear syntactic structure, have slightly more difficulty with this sentence than with a sentence like 'The fence was kicked by the cow,' which is easier because fences don't kick, so it is obvious that it was the cow that did the kicking. Our language here is subtly affecting our thinking, or at least our thinking about language.

But with effort and with time to think, we can display cognitive independence from language. French speakers can do as well as English speakers in translation, reading, and so on about manner of action, if they are given more time and asked to think about it. Although the language influences them, it does not by any means exert a tyrannical control over their thoughts.

Many researchers join Slobin in accepting this weaker version of Whorf's hypothesis. One prominent group is led by Dr Stephen C. Levinson of the Max Planck Institute for Psycholinguistics. Levinson's research group is large and well-funded and they have made some truly impressive discoveries. One that I particularly like, because it relates to some of my own observations among the Pirahãs, has to do with how people navigate through the world around them, how they orient themselves spatially to their surroundings.

There are many groups in the world who, like the Pirahãs, do not use their bodies to orient themselves directionally, but, rather, use the world outside their body for this purpose. The Pirahãs would never say 'turn left' or 'turn right' but 'turn up river' or 'turn towards the jungle' or 'turn down river,' and so on. The average American, however, does not use such a system, relying on their bodies to give and take directions. If an American were to give you directions, they would most likely tell you things like 'Then you turn right. Right after that you turn left. Then you want to proceed ahead for a while. If you go too far, back up.' These are all body part references, based on left hand, right hand, head, and back. They work well if the one taking the instructions can keep straight in their head which way the direction-giver is facing or which way the direction-giver expects them to be facing. If I am facing north, then a left turn is west. But if I am facing south, a left turn is east. It would have been less ambiguous to say, 'Turn north. Then follow this with a quick jog west. Then another quick jog, this time east, and you can proceed.' But even though English speakers can use non-body-relative terms like north, south, east, west, etc., most of us do not. We do not even know most of the time where north, south, east, and west are. By contrast, the Pirahãs do not, and *cannot*, tell someone to turn left; instead they use external geography.

What Levinson's research group have shown is that people who come from Pirahã-like directional and spatial systems are better able to navigate in new environments than people like us who primarily rely on the body to orient ourselves in the world. The conclusion his team draws from this work is that language has an effect on our thinking – relying on directional language that is based on the external world helps us to think more effectively about navigation. Although Levinson's research has been criticized, it does mesh with my own, informal observations about the Pirahãs (which could, I admit, be skewed because I am so bad at finding my way around and so am deeply impressed with even moderate ability in this area).

Steven Pinker criticizes the Whorf hypothesis in his influential 1994 book *The Language Instinct*. Although Pinker is careful to distinguish between linguistic determinism and linguistic relativity, he says that Whorf's hypothesis and predictions are 'wrong, all wrong.'

The fair assessment is that the way we talk sometimes affects the way we think. But also, the way we live culturally affects the way we think, too. We can usually think independently of language and culture if we take our time, though it can be very difficult to do so. And always our thinking is limited by our biology. There is no simple or uncontroversial theory about the interaction between language and thought. Life, language, and thought have a complex inter-relationship. Answers will not always be neat.

As we saw, Pirahã color words and kinship terms offer insights into a cognition that contrasts sharply with all versions of Whorfianism. Thinking in these categories does not always require language. And yet what we learn by all of this is that words are as much tools as pliers and pipe wrenches. Just as we use pipe wrenches for plumbing and alcohol for relaxation, we use number words for math – if we don't have the words we cannot do the math. Math is a cultural tool itself that relies on cultural-linguistic tools. If a society wants math, it will need number words. It can invent them or borrow them from another language, but it will have to get some. Otherwise, math cannot be done – the tools to build that edifice would be lacking.

The Pirahã people can do some things that involve quantities and numerical relationships without using number word tools. For instance, they regularly order things: what to do first, what to do next or last; who arrived in what order; who was born first; the order in which we will find villages descending or ascending the Maici river, and so on.

If you do not have the correct language tools to hand, though, you can still express things by using improvised, poorly designed tools. For example, I could find out if you were older than your sibling by asking you who was born first. The literal question I would ask is, 'Who was expelled from the stomach *at the head*?' (**Kaoí xapaí kaopá**), the phrase 'at the head' implying 'first.' Or I could ask, 'Who was expelled from the stomach *at the butt*,' in which 'at the butt' (**Kaoí tiapaó kaopá**) implies 'last.'

If you have multiple siblings you could tell someone about everyone's birth order by saying 'John was expelled at the head. Then Peter. Then Mary. Then Joe. Sue was at the butt.' And there you have a perfectly clear ordering relationship without numbers. But we can see that, if the list were very large, it would take much more concentration or mental effort to keep track of all the items on the list without ordinal numbers like 'first,' 'second,' 'third,' 'fourth,' and so on. It is not impossible to do this but it is unquestionably more difficult. But say that you want to use an ordinal number in another way, such as 'Stop at the fifth house on the third road.' Well, if you are a Pirahã speaker, now you are out of luck. There is no way to communicate this idea in Pirahã.

Without the tools of ordinal number words, there are things that the Pirahãs cannot talk about. But of course, if you know everyone in your society and they all live along the same river's edge, you are never going to need to say things like 'the fifth house on the third road.' The ordinal number tool is not needed by the Pirahãs. If Pirahã culture radically changed, however, and suddenly there was an acute need for ordinal numbers, then presumably the Pirahãs would respond just as all cultures without ordinal numbers have responded to that need throughout human history – they would invent them or borrow them from another language that already had them.

We see this in language and science all the time. When Murray Gell-Mann needed a new term for the elementary particles of nature he had discovered, he coined the term 'quark,' based on the sound the ducks made in James Joyce's 1939 fiction *Finnegans Wake*: 'Three quarks for Muster Mark! Sure he hasn't got much of a bark. And sure any he has it's all beside the mark.'

Although it is remarkable that the Pirahãs lack number words and although this shows that words are tools, this does not demonstrate that grammar is a tool, governed and created by culture like any other tool. We need to find evidence that grammars themselves are tools.

The phrases 'cognitive technology' and 'cognitive tools' have become popular in recent years, as researchers have begun to reach consensus that the idea that the utilitarian aspects of language and its contribution to solving problems of human societies, whether the learned or the invented, should be at the center of our explanations of the nature of language. Although Vygotsky's influence among linguists is still minimal, his pioneering ideas are behind much of these recent moves toward tool-based explanations of cognitive abilities.

I began to think in these terms in the early 1990s, while writing a paper on the relative lack of words to express precise times and tenses in Pirahã. I first made the case in a 1993 paper in which I showed that Pirahã has no perfect tense and I provided a theoretical means for accounting for this fact. A perfect tense in English would be 'had eaten' in 'I had already eaten by the time John arrived' or 'will have eaten' in 'I will have eaten by the time John arrives.' Perfect tenses differ from other tenses because they indirectly refer to the present, the past or the future, although the time they actually refer to is fixed by another verb in the sentence. This would be 'arrived' with 'had eaten' and 'arrives' with 'will have eaten.' The Pirahãs lack this kind of tense because all their references to time are relative to the present, not to hypothetical events in the past or the future.

The absence of Pirahã perfect tenses indicates not merely the absence of a special tense word or suffix, but a much deeper cultural lacuna. There is *no* way to convey a perfect tense meaning ever in Pirahã. In fact, Pirahã has very few words for time, period. The complete list of Pirahã time words is: **xahoapió** 'another day' (literally, 'other at fire'), **pixí** 'now', **soxóá** 'already' (literally, 'time-wear'), **hoa** 'day' (literally, 'fire'), **xahoái** 'night' (literally, 'be at fire'), **piáiiso** 'low water' (literally, 'water skinny temporal'), **piibigaíso** 'high water' (literally, 'water thick temporal'), **kahaixaíi xogiíso** 'full moon' (literally, 'moon big temporal'), **hisó** 'during the day' (literally, 'in sun'), **hisóxogiái** 'noon' (literally, 'in sun big be'), **hibigíbagáxáiso** 'sunset/sunrise' (literally, 'he touch comes be temporal'), **xahoakohoaihio** 'early morning, before sunrise' (literally, 'at fire inside eat go').

I wrote a report of my analysis of Pirahã time words and tenses and, like any other researcher, circulated a draft of my paper to some respected colleagues for comments before sending it to a journal for more rigorous peer review. One of these colleagues commented drolly that it was not terribly surprising, after all, that a group of folks living such a temporally homogenous existence would turn out to lack words that differentiated periods of time that were not important to them in the first place.

My colleague's remarks made me think deeply about this. First, they made sense. Second, I knew that there were counterexamples in other languages and cultures – where life was similar to the Pirahãs', but where tenses were abundant. So I would have to think about this carefully. The match between Pirahã culture and its relative dearth of time expressions could be merely a coincidence. But there is no controversy to the assertion that the Pirahãs do not need a wide array of time words. These words have no work to do in a society in which members sleep, eat, hunt, fish, and gather, without regard for the time of day, day of the week, week of the month, or month of the year. This insight led me to take a fresh look at other areas of the Pirahã language where there might be a culture–language connection.

Our assumptions drive our questions. If we believe that all cultures and languages have numbers, we are unlikely to test to see if numbers are present or absent. If we believe that all languages have roughly identical grammars, we are less likely to probe for differences, or at least for profound differences. And yet, the quandary is that we need to have beliefs and theories about the nature of the object we are studying, whether it is a language or a cancer, if we are to ask any well-formed questions. There are two ways out of this quandary for me. First, our theories of language must include culture. Second, we should never take our theories *too* seriously.

If you have a theory of language or grammar that begins with the assumption that culture plays no role, then you are not likely to notice a role for culture.

The lack of certain vocabularies like numbers in a particular group can teach us several things, however. First, if we discover that a category of meaning previously believed to be universal, such as numbers, is not found in all languages, then this is itself an important result. It teaches us that, for whatever reason, the presence of this category does not define our species, that you can be a normal human being without it. Second, the relative ability of people to think about a concept, even when they lack the vocabulary for it, can teach us something about the nature of the concept and the effect language does or does not have on concept formation.

The absence of number and color words in Pirahã helps us grasp why, if languages are cultural tools, certain kinds of words and expressions will be lacking in some languages. But there is further evidence for things that are unique to Pirahã, or very rare, which offer positive support for the use of linguistic tools by this language. These tools are particularly interesting here because it is their utility that explains their existence. Pirahã channels of discourse are one such positive example of why language should be understood as a cultural tool.

Pirahã is a tone language. This means that pitch – that is, the controlled frequency of vibration of the speaker's vocal cords – is very important to Pirahã communication. The Pirahãs exploit their ability to perceive tones in their language to develop options for communication that are unavailable to speakers of most European languages.

I experienced these channels for the first time one afternoon after I had set out some old National Geographic magazines for the Pirahãs to thumb through. A woman, Xiooitaóhoagí (i-owi-taO-hoa-gI), sat on the floor looking through the magazines, nursing her baby, her legs sticking out directly in front of her and her dress pulled down to her knees, in the normal Pirahã manner. As she looked, she hummed to her infant. And as she hummed I came to realize that what she was humming was a description of the whale and Inuit whose pictures she was examining. The boy would look away from her breast to the picture from time to time, and she would point and hum louder.

With hum speech, Pirahãs can say anything that can be said with consonants and vowels. Hum speech is used to disguise either what one is saying or one's identity or, as in the example here, for mothers or other caregivers to communicate with their children. Even for a native Pirahã hum speech can be hard to follow if they are not paying close attention.

A second channel of discourse in Pirahã is yell speech. This channel is created by using the vowel 'a' or, occasionally, the original vowels of the words being used, and one of the two consonants 'k' or glottal stop, to yell to someone far away or during a lot of noise, such as a thunderstorm. The most salient features perceived at a distance are the tones.

And then there is musical speech, one of the two communication channels that the Pirahãs have special names for. They refer to musical speech as 'jaw going' or 'jaw leaving.' It is produced by exaggerating the relative pitch differences between high tone and low tone and changing the rhythm of words and phrases to produce something like a melody. Musical speech has an interesting array of functions in Pirahã, two of the most important of which are to communicate important new information and to communicate with spirits.

The final channel is whistle speech, which the Pirahãs refer to as talking with a 'sour mouth.' It is used to communicate while hunting and in aggressive play between boys. Only males use whistle speech.

These channels of discourse are tools, each with a form that fits its functions well. They are creations of the Pirahãs' culture and language. One must use the appropriate channel of discourse in the appropriate way to communicate the different types of information that each of these channels was created to transmit.

The frequency with which these channels are used is determined by Pirahã culture. This is because they *choose* to define the ways in which the channels should be used and when to use them. These social choices are determined by cultural values. Therefore, in order to fully understand these channels, their use, nature and origin, it is necessary to understand Pirahã culture.

We have seen how words, culture, and cognition can be related in different ways by the examination of kinship terms, color words, and math in Pirahã. There are many linguists, however, who will continue believe that culture's effect on language is insignificant unless we can show it affecting the way syntax – the core of a language's grammar – works. So that is where we are headed next.

## Chapter Eleven

# YOU DRINK. YOU DRIVE. YOU GO TO JAIL.
## CULTURAL EFFECTS ON GRAMMAR

*'How complex or simple a structure is depends critically upon the way in which we describe it.'*

Herbert Simon, *The Sciences of the Artificial* (1969)

O ur grammars follow our values. This is especially clear in the ways in which literacy changes the grammar of its readers. There are many examples of literacy's power to change in history but my favorite comes from the story behind the Cherokee writing system. This is an impressive example of genius that illustrates the point that we develop cognitive tools as we need them for socio-cultural reasons.

In all of human history, writing has only been invented a handful of times, in Guatemala, in Sumeria, on Easter Island, in Egypt, in China, and in a very few other places. One of these rare moments in the story of our species occurred in 1821, as the result of the keen mind of a crippled Cherokee-American silversmith named George Gist in English and Sequoya ('Hog') in Cherokee. His mother was Cherokee and his father was likely the wealthy American fur trader Nathaniel Gist. Although Sequoya fought alongside the Cherokee against other Indians, he also spoke English and made a reasonable living at his trade as a silversmith, living among non-Cherokees.

Some Native Americans were impressed by how 'whites' used writing in order to communicate with others across space and time. Sequoya was more than impressed – he was challenged to create such a system for his people.

He developed an orthography unlike the alphabetic system used by Europeans or the ideographic system used for Chinese. In Sequoya's

| | | | | | |
|---|---|---|---|---|---|
| Ꭰ a | Ꭱ e | Ꭲ i | Ꭳ o | Ꭴ u | Ꭵ v |
| Ꭶ ga Ꭷ ke | Ꭸ ge | Ꭹ gi | Ꭺ go | Ꭻ gu | Ꭼ gv |
| Ꭽ ha | Ꭾ he | Ꭿ hi | Ꮀ ho | Ꮁ hu | Ꮂ hv |
| Ꮃ la | Ꮄ le | Ꮅ li | Ꮆ lo | Ꮇ lu | Ꮈ lv |
| Ꮉ ma | Ꮊ me | Ꮋ mi | Ꮌ mo | Ꮍ mu | mv |
| Ꮎ na Ꮏ hne Ꮐ neh | Ꮑ ne | Ꮒ ni | Ꮓ no | Ꮔ nu | Ꮕ nv |
| Ꮖ qua | Ꮗ que | Ꮘ qui | Ꮙ quo | Ꮚ quu | Ꮛ qvv |
| Ꮝ s Ꮜ se | Ꮞ s | Ꮟ s | Ꮠ s | Ꮡ s | Ꮢ s |
| Ꮣ sa Ꮤ te | Ꮥ se Ꮦ te | Ꮧ si Ꮨ ti | Ꮩ so | Ꮪ su | Ꮫ sv |
| Ꮬ dla Ꮭ tle | Ꮮ dle | Ꮯ dli | Ꮰ dlo | Ꮱ dlu | Ꮲ dlv |
| Ꮳ tsa | Ꮴ tse | Ꮵ tsi | Ꮶ tso | Ꮷ tsu | Ꮸ tsv |
| Ꮹ wa | Ꮺ we | Ꮻ wi | Ꮼ wo | Ꮽ wu | Ꮾ wv |
| Ꮿ ya | Ᏸ ye | Ᏹ yi | Ᏺ yo | Ᏻ yu | Ᏼ yv |

Figure 10 **Cherokee Syllabary invented by Sequoia**

system, a symbol was created for *each syllable* of Cherokee. Thus, instead of an alphabet, he invented a syllabary. The symbols that Sequoya developed and the syllables they represent are given in the chart below:*

Each syllable, usually a consonant followed by a vowel, is represented as a single symbol in the writing system. For languages like Cherokee with a relatively small number of syllables this is an efficient way to learn and teach writing. Other languages that have used syllabaries include Mycenaean Greek, Japanese Katakana and Hiragana writing, and Inuktitut.

Syllabaries, as it turns out, are *for some languages* superior to the alphabetic writing systems that use one symbol per speech sound that we find in English, Spanish and most other written languages. For example, syllabaries are claimed to be more intuitive for the learner and thus easier

* More on the Cherokee syllabary and other writing systems of the world can be found at http://www.omniglot.com, from which the chart of the syllabary is taken.

to learn. The claim here is that people in some languages think in terms of syllables when they reflect on the pronunciation of words more than they think of individual sounds. Effective syllabaries have fewer symbols and rules for combing symbols than many alphabets.* In spite of its advantages, however, the Cherokee syllabary was ultimately replaced by an English-like alphabetic orthography. The reason that this move occurred is that, even though the syllabary might have been the best way to write Cherokee, the people who learned it would have to learn to read English or another language as a completely new process, since both the symbols used and the forms they chose to represent (syllables or individual sounds) are so very different between English and Sequoya's system. This shows us that even a good tool, such as the syllabary of Cherokee, will fall into disuse if a tool that is better for new tasks – in this case, Cherokee alphabetic writing – becomes available. It shows that tools are designed to fit tasks such as to read only Cherokee (the syllabary) or to read Cherokee in a way that the skill transfers to English (alphabet).

Not only did Sequoya not imitate the European writing system (though he was influenced by it, of course), he improved on the invention for the purpose of reading Cherokee. And he invented all of the symbols used. He had to invent them all because there was no equivalent to the Cherokee syllables he could have borrowed from the English alphabet. Sequoya's syllabary is a work of first-order genius, even though his brilliance is rarely recognized.

One institution that publicly acknowledges the work of Sequoya is the Carnegie Museum in Pittsburgh, facing the University of Pittsburgh's famous Cathedral of Learning. Having entered the Carnegie, walk straight ahead through some impressive marble columns. You will enter a large room full of replica Greek statues and other art of

---

* However, a syllabary would not work well for English, as linguist Chris Barker has shown. Barker calculates that English has 15,831 possible syllables that would need to be represented in a syllabary. That is not a good beginning for an English syllabary, since English has only twenty-six letters in its alphabet (even though many English dialects have upwards of forty actual speech sounds, several of which are not represented in the English alphabet). Barker's work underscores the fact that the forms of language, written or otherwise, are tools and that tools must fit their tasks.

the ancient world. Next comes the good part. From the center of this classics room, look straight up at the ceiling. Here you will see what appear to be frames of frosted glass, each frame containing a symbol in it. These symbols, which people rarely look up to see and which they almost never recognize when they do, are the syllabary of Sequoya.

Sequoya's invention did more than just provide a new cognitive tool for the Cherokee. True, this tool enabled them to join other peoples with written forms of their language in being able to store thoughts semi-permanently outside their bodies. But what was more important was that this tool changed the Cherokee language and culture. Like the infinite regress of one mirror held up to another, a culture affects its language which affects its culture which affects its language, and so on ...

Writing changes a language. It does this initially, of course, in the written language itself. But its influence extends into the spoken language as well. The language's sentences become more complex. Its readers are transformed from mere receivers of information to editors, thinking of better ways to express a thought. They have time to concentrate on their language. And literacy does more to a language than to produce longer sentences. Written words can be chosen with greater precision. This is why we almost always hold people more accountable for their written words than for their informal spoken remarks.

Written discourse can also be longer and much more complicated than spoken discourse. For example, people rarely utter words in the same number or complexity as an author would do in writing a novel. The writer becomes an editor, not merely a transmitter. Writers spend more time forming their stories, speeches, sermons, and sentences than other people. A reader can read for a while, get up, have dinner, and then return to their reading, or even reread. They therefore have much more time, at least potentially, to process what they are reading than if they were listening to someone speaking. Reading can be done at your own speed, unlike listening or speaking.

Many linguists, anthropologists, and historians have observed changes in languages whose oral–aural channel has been supplemented by a graphic channel. Listen to the recorded folktales of a people. Then compare the cadences, complexity, and consequences of their written stories to these oral tales. When we learn to write down our language, the reason we can make our sentences longer and more complex is because the graphic medium is free from noise and can be left and returned to at will. Our language becomes visual in a new sense, beyond gestures and body language, and we consequently think of it in a new way. How many people can make a living in a modern industrial culture as a 'story-teller' by simply speaking? Not many. Yet many people write stories for a living rather than tell them.

The alphabet is a culturally invented cognitive tool. The internet is also a cultural tool. The former is perhaps more basic. But a tool is a tool and I am happy to use them all, as I need them. There is no escaping tool use. And who would want to escape? Humans may not be the only species to use tools, but we have the best ones. No other species has anything approaching our brains and the networks – cultures and societies – to which they are connected. We are each a single light on a worldwide power grid.

Our cultures are our greatest non-biological tools. Culture is how a pre-graphic Amazonian survives. Just as Americans survive. By cunning and cooperation, by individual effort and ability multiplied and enhanced by the connected minds of his family, friends, and fellow community members. How does a Londoner survive? A member of Al Qaeda? How does a clarinetist succeed playing a symphony in the New York Philharmonic? Exactly the same way. Could one person alone perform a full orchestral piece live, without electronics? Hardly. Could one person be an economy? Not even Carlos Slim.

Sequoya showed that it is cultures that produce graphic language out of need. As I have mentioned in chapter six, the principal difference between spoken and later written forms of language is the fact that sounds do not hang around very long in spoken language; that is, they are subject to rapid fading. When you say a word or sing a song, all of the sounds you produced disappear almost immediately. There is no instant replay for speech.

The phenomenon of rapid fading affects the grammars of human languages. For example, one of the most noticeable characteristics of exclusively oral–aural languages, such as Pirahã, is redundancy. We hear frequent repetitions, both in naturally occurring conversations and stories. Redundancy is necessary because the failure to hear rapidly fading words in a noisy public environment means that conversations would be harder to follow without some repetition. In a Pirahã story, for example, it is common to hear lines like 'so and so said …' repeated three or four times in a row.

Here is an example of what I mean. It is a story told to me by my good friend Xahoápati, about his only son, Xigábaí. Xahoápati told this story excitedly, because he was emotional from almost having lost his son.

### Xigábaí almost bitten by snake
Told by Xahoápati to Dan Everett
*c.* 2000

1. **Xaí ti igaí apaí.**
'So I spoke at first.'

2. **Xigábaí gí basí igabopapí.**
' Xigábaí was bringing your bed back up.'

3. **Xís ibáitaaboihaítigaiti.**
'He arrowed the snake a couple of times.'

4. **Xís axáahá.**
'The snake bit [at him].'

5. **Xaí Toítoí hi xigíaxáaihá.**
'Thus Toítoí with [him] it bit at [also].'

6. **Hi agíai tigaiti gíi xibigaá.**
'Thus it was a real snake, a bushmaster, they saw.'

7. **Tigaiti gíi xaí. Xis ibáitabogaáti.**
'It is a bushmaster. Shoot it!'

8. **Hi agía xogi áagaó xis aihi ogaáabópapá. Xaí tigaiti.**
'Thus the bulk of the people wanted to search carefully down on the ground.

Thus [for that] snake.'

9. **Ti agía xis igíokáabópápáxaí.**
'[When I was] with the animal, [it] struck up [at me].'

10. **Ti igáísai. xi abáopísaxáagabagaá.**
'I spoke [as I have been doing]. It almost bit him, almost wanted to.'

11. **Xiabáísi káaboópaihiabisóai. Xáihá.**
'Thus, it [the bushmaster] didn't bite the human being [upwards and viciously].'

12. **Xi goó ixaóopáahaxaí.**
'He was there [almost] bitten?' [It was there that he was almost bitten.]

13. **Xi goó ixaóopá. Xoí hi aí ixíga.**
'He was there [almost] bitten. It was there in the jungle.'

14. **Xigábaí hi aabáobíisaxáabaí.**
'[The snake] almost bit Xigábaí, as we have been saying.'

15. **Xigábaí hi áaboópáihiabsóixáihi.**
'[The snake] barely didn't bite Xigábaí.'

The snake striking at Xigábaí is the theme of this story. Because of rapid fading (in a noisy oral–aural environment) and to emphasize its importance, the striking and the almost getting bit events are repeated several times.

Now contrast this with a typical English newspaper or magazine story. They are not subject to rapid fading, and printed words are expensive. Therefore a written text takes on a different appearance than an oral story. Take, for example, the following excerpt from the *Chronicle of Higher Education*:

*Students from Lafayette College are cultivating new cash crops with subsistence farmers in Honduras, working with hurricane-devastated residents of New Orleans to attract businesses to the area, and drawing up plans to encourage arts-related tourism in their college town.*

*Such economic and civic engagement would be commonplace at land-grant universities and engineering schools. But it is far less familiar at liberal-arts institutions like Lafayette, which has embraced such work as a way to connect to outside constituencies and to demonstrate the value of a liberal-arts education.*

There is little repetition in this excerpt, even though it is bristling with new information and requires a good deal of cultural background for interpretation. Each sentence in it is longer, structurally more complex, and more packed with new information than any of the sentences in the Pirahã story. These differences are heavily influenced by the different functions and capabilities of oral discourse on the one hand – at the mercy of rapid fading – and written discourse, which is in principle around forever, on the other.

To the outsider, the repetition in Pirahã discourse can almost sound like stuttering. But because speech fades away so quickly and because the environment is noisy (crying babies, laughing children, squealing adolescents, dogs barking, and so on), the listener needs help to avoid losing the train of thought because someone, say, sneezed just as the speaker said something important. Redundancy compensates for environmental noise. The same short life of sounds keeps the sentences of most speakers brief – the longer the utterance the greater the chance that the hearer will be distracted or noise will cause a break in the transmission of the message.

There are commonly other aspects of small societies using the oral channel of discourse exclusively. These can vary from group to group and perhaps indicate that a new subfield of 'hunter-gatherer linguistics' might be worth exploring.

A common characteristic of human language is the property of recursion, which, as we have seen, is the nesting of one sentence inside another. I do not think that anyone disputes that recursion is universal among humans, though where it is to be found has been a matter of hot debate. After all, normal humans must be able to think that other humans think. And they must be able to think that other people know that other people think. That is one thought inside another thought of the same type – recursive thinking. And they often want to tell other humans about their thoughts of other people's thoughts. Suppose you think that 'Mary knows that I know that her husband knows that we know that he is fooling around.' That is recursive thinking. Without it we could not have such thoughts. But it turns out that where the recursion is located in our languages – in our stories, in our sentences, or both – has very little to do with our ability to reason recursively.

Think of the 'fooling around' sentence above. There are four thoughts here. Except for 'Mary knows,' each thought is contained in another thought. What must our languages be like to report such matrushka doll-like thinking to another speaker? This kind of engineering problem for language shows us that language is quite complicated. In the engine room of language, there are cogs and wheels that enable us to make big meanings out of little meanings. Much of modern linguistics is about figuring out the workings of the complex cognitive engine that enables us to utter complex thoughts in a form that can be nearly instantly grasped by others.

But is the linguistic engine room the same for all languages? Or could it be that some languages have a straight eight, others a V-4, and still others a Wankel engine? These are different solutions to powering an automobile and each has its advantages and disadvantages. And answering such questions about how languages function requires us to get our hands greasy. We have to talk a bit about some of the design issues and challenges that produce the engine rooms of languages.

These design questions get us back to Hauser, Chomsky, and Fitch's proposal that all languages are based on recursion. Although these authors do not define recursion, the basic idea is simple. The structures produced by recursion are structures in which one unit is found within another unit of the same type. Clear enough. However, if we focus on the *process* of recursion, rather than its output, then a recursive operation is said to be one that takes its own output as the input for its next operation. This sounds more complicated than it is.

For example, take a noun and make it part of another noun. You might start with a noun like 'room' and add that to another noun to produce, say, 'bathroom' – though written as two words in the English orthography, it is just a single noun grammatically. Or take a sentence, such as 'John went to town' and place it in another sentence, 'Bill thinks,' to get the larger sentence 'Bill thinks John went to town.' Or take the prepositional phrase 'under the bed' and place it, as a modifier of 'box,' within the larger prepositional phrase 'into the box under the bed.' All humans have the ability to take language units of one type – words, phrases, sentences, thoughts, paragraphs, and so on – and place them inside another unit of the same type.

If grammarians or computer scientists write rules to form one unit out of another of the same type then the grammar those rules constitute is recursive. Most theories of grammars are recursive in this sense. But we do not need grammars to illustrate recursion. It is all around us. If you hold up a mirror in front of another mirror, the reflection of one is picked up in the reflection of the other and produces an 'infinite regress,' a never-ending set of reflections. When Jimi Hendrix held his Fender guitar in front of his Marshall amplifier at Monterey in 1967 and the Marshall amplified itself amplifying the guitar in an infinite aural regress, the effect was the high-pitched squeal that is the signature of auditory recursion known as 'feedback.'

But for recursion to be part of humans' unique language endowment, as Hauser and company propose, it would be necessary to show that animals are incapable of recursive communication or thought. On the other hand, if animals *were* capable of recursive thought, we would be forced to conclude that human language draws upon abilities that humans share with other creatures. The uniqueness of human language would therefore lie not in recursion but in the way in which *Homo sapiens'* brains have been able to exploit other components supplied by evolution to a number of species.

In 1992, psychologist Irene Pepperberg published an article in the *Journal of Comparative Psychology* in which she presented evidence that an African gray parrot called Alex could master tasks involving conjunctive reasoning. In which case it would appear that at least one African gray parrot was capable of recursive thought. Pepperberg's article received very little attention from linguists when it was published, because at the time recursion was not the popular topic it has since become. But if Pepperberg is correct, recursion is not specific to humans and therefore it cannot be the single element that makes human speech possible. And this wasn't all Alex could do. According to numerous reputable sources, he was capable of taking turns in conversation with Dr Pepperberg, a feature crucial to human conversation; he had a 150-word vocabulary; he could sort and label; and he could count, among many other things.

Although these claims are all controversial, most of them emerge from interpreting the data of rigorously controlled, scientific tests and cannot therefore be ignored. We of course need more data from Alex and other parrots and animals. Alas, Alex went to parrot heaven in 2007 (his last words to Pepperberg were 'You be good. I love you'), so it is not possible to replicate the experiments with him. Someone needs to conduct experiments with other gray parrots and animals to look for evidence of recursive reasoning elsewhere in the animal kingdom. In the meantime, the studies of Alex the parrot should temper our views of what is 'unique' to humans.

Dog-barking is another candidate for recursive behavior. When a dog barks more than once, then we could describe this as barking within barking – recursive barking. Or try calling your cat to you. Let us say that it starts towards you but then on the way it runs over to get a drink. Then it continues on towards you. Should we describe the cat's entire path, one that contains both a 'drink path' and a 'go to (the person who thinks they are your) master' path within it as a recursive path? If we do, then we are claiming that cats and other animals construct recursive paths. Moreover, if these recursive descriptions are accurate, then recursion clearly is not unique to humans.

Of course, I recognize that for these arguments to be valid in the sense of recursion that Hauser, Chomsky, and Fitch had in mind, we would need to show that there was a recursive process in the minds of these animals. Some might argue that this claim for 'recursion' should be more than merely the way I choose to describe what the animal is doing. But that would be to raise the standard of analysis higher than we place it for humans. When someone claims that a human language has recursion, what they mean is that this is the best way for the grammatical theory they are working with to describe the language.

Having said all that, when my dog barks, it is possible that he really does intend to bark a lot. Is this entire barking behavior the result of a dog constructing a single event of 'barking' out of individual barking productions? If so, the barking is recursive and we can say that it has a structure like *[*[bark][bark][bark][bark][bark][bark]*]* – a lot of barks contained in one larger barking event. Or is the barking just a separate set of barking behaviors –[bark][bark][bark][bark] – each independent of the other? The same evidence that would lead me to describe a human as intending to do something would lead me to describe my dog as intending to bark and intending to bark multiple times. If that were accurate, my dog is capable of recursive mental operations. And, getting back to that cat, how do we determine whether its path from where it began to its drinking bowl and then to me was the result of a single operation – one large path with two subpaths – or just two paths, two separate events? Is the structure of its path the result of recursive thought – (Go to master by way of getting a drink) – or not – (Get drink) (Go to master)? If it intended both to get a drink and to come to me, then this means that the cat has coordinated – and hence recursive – intentions. (This is because coordination, such as 'John and Jill,' 'Do this and then do that,' is recursive. You take one noun phrase, for example, 'John,' and make a larger noun phrase containing it, such as 'John and Jill.') And coordination is recursion. This is all speculative. But it is not obviously false. In fact philosophers, in work on intentional cognitive states like beliefs and desires, have readily attributed intentional states to other animals.

Imagine the following exchange. You're back at school and your teacher mumbles something to you about putting the pencils you have been using in class away in some box. You do not understand the teacher well, so you ask the question, 'Put the pencils into which box?' The teacher responds, 'Put the pencils into the box under the bed.' The phrase 'into the box under the bed' has one prepositional phrase inside another and is therefore recursive. We can diagram this recursive prepositional phrase in a linguistic 'tree diagram', similar in spirit to a Reed-Kellogg diagram, but intended to be applicable to any language in the world.

## Figure 11 **Recursive Prepositional Phrase**

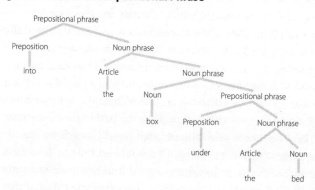

Recursion becomes very important in our discussion at this point because of the way that it is used, constraining entire grammars and the genius of a whole language. In some theories, tree diagrams like the one in Figure 11 are believed to 'model' the knowledge that speakers have of their languages. In Chomsky's theory of grammar, which informs his work with Hauser and Fitch, such trees are always built by recursive application of syntactic rules.* There is no way to produce an utterance in Chomsky's theory other than by recursion. Thus it is only natural that the paper that Chomsky and his co-authors produced would place a high priority on recursion.

But although all linguists recognize the occasional usefulness of recursion in sentences, they do not all accord it the same theoretical importance. For many philosophers, linguists, anthropologists, and psychologists, what really distinguishes the communicative ability of our species from communication in other creatures is that we can tell stories. Ironically, stories do not count as part of the grammar to a large number of theoretical linguists. Thus structural patterns discovered in discourse cannot be used by the practitioners of such theories, so facts

---

* In Chomsky's latest theory, which he refers to as minimalism, he proposes that speakers of all languages use only one rule to produce their grammatical structures – words, sentences, phrases, and so on. It is a recursive operation which he refers to as 'Merge.' This is why recursion is so vital to his theory of language – without it his theory does not work.

about stories are rendered irrelevant to their claims about the nature of grammar. And it is important to remember that when some researchers talk about 'language' they almost always mean 'grammar.' In fact, they mean only a special slice of grammar called 'formal grammar.'

Recursion is an important concept to grasp because it helps us understand why humans can talk endlessly about something, adding more and more details, clarification, side comments, and so on – in theory, for ever. A story could in principle be taken up without interruption by subsequent generations. So it is not surprising that the famous nineteenth-century German explorer and polymath Wilhelm von Humboldt referred to language as the 'infinite use of finite means,' a statement cited frequently by linguists. This infinitude of human languages is often claimed to be one of the most important things any theory about language has to express. But where does this infinite use reside? According to some, in recursion in sentences. Now, I agree that recursion is crucial, but I do not think that sentence recursion *per se* is all that important. Rather recursive *thinking* is what is crucial. One way to approach this issue is to think of recursion in thought and language as tools, but that recursion in language shows up when and where a culture desires it, if it does at all.

If recursion were found to be lacking in the sentences, phrases, and words of a specific language for cultural reasons, two very important questions would have to be answered. First, what would be the implications of this lack for the cognitive functioning of the community speaking the language? And secondly, what cultural mechanism could so police the grammar as to ensure that no one talks recursively?

That is the question that must be asked if we are going to convince linguists that they need to incorporate the effects of culture in their grammars. But before turning to that I want to emphasize that finding evidence that recursion or some other grammatical characteristic is missing or present in a given language tells us nothing about the overall complexity or worthiness of that language. Languages incorporate what they need to, according to the demands of their cultural niche. As the niche changes, the language can also change. This is because the source of language is the culture and psychology of the speakers. Nevertheless, if there is evidence that something is absent from a given culture or language, that tells us something interesting about that culture or language and about humans in general.

Looking at the Pirahãs' language, one is impressed by the fact that every man, woman, and child seems to have mastered the knowledge of the world that surrounds them, in ways that few western cultures would be able to do. The Pirahãs have and use words for every bit of flora and fauna in the air, river, and jungle they inhabit. These words respond to Pirahã cultural needs – especially, the need to understand the environment that they must master to survive. The rich and detailed knowledge of the average Pirahã of this environment is remarkable. But just as the Pirahãs invent and master words for the things that they do need, they, like the average speaker of any language, lack words for things that they do not need, even words for things that are acutely needed by other societies, such as numbers.

Getting back to the issue of recursion, there are many words in English that tend to be found in sentences with recursion. English words like 'think,' 'believe,' 'say,' and 'want' all require subordinate clauses. We see this in examples like 'John said that he was coming,' where 'you are coming' is a subordinate clause. An English speaker cannot say 'I think you' without the following clause, because that is just the way the verb 'think' is programmed in English.

Pirahã lacks all words of this type. No verbs in Pirahã require a subordinate clause. This is one way that Pirahã culture can eliminate, by lexical convention, recursion from its grammar. Rather than saying 'John said that he was coming,' the Pirahãs would say 'John spoke. He is coming.' That is ambiguous, making what is in English a single sentence in effect a small story. Or rather than say, 'John thinks he is coming,' they would handle this the same way, 'John spoke. I am coming.' Like other Amazonian groups, the Pirahãs tell us what they think that people are thinking or what they themselves are thinking, by using the verb 'to speak,' putting words in other people's mouths to convey their thoughts.

Alternatively, a language could eliminate recursion from its sentences by restricting recursion to stories, developing recursive relationships *between*, rather than within sentences. It seems that all people not only reason recursively, but that they need recursive language as a tool to express recursive thoughts. However, if the recursive thoughts a culture wishes to express can be expressed adequately without recursion in the syntax, then the tool of recursion are unneeded outside of discourses. This would be reinforced if there were other cultural principles that made recursion less desirable for that language, as has been claimed in several publications about Pirahã.

The grammar of the Pirahã language is famous because it appears to lack recursion in its sentences and phrases (even though it has recursion in its stories) and because it has been claimed this lack of recursion is forced by the culture. Since claims were made to this effect in 2005, critics have played a useful role in helping to refine, rethink, and strengthen the case. One of the many pleasing aspects of science is that colleagues, even as they disagree vehemently, can lead one to better thinking and to helping one decide whether to abandon an idea or to try and express it more clearly.

Pirahã is very different from many other languages. In most grammars, phrases and sentences are complicated enough to require description by the tree diagrams in Figure 11. But in Pirahã much of the linguistic complexity is found in stories, rather than in sentences.

Since this is not a grammar book, there is no need to go deeply into the details of Pirahã sentence structure. The crucial matter is to actually give an explicit analysis of how a culture can affect a grammar, that is, the very structure of its sentences.

There are at least two controversial aspects of my analysis of Pirahã. This first is that if culture can constrain grammar, then grammar is not prespecified in some instinct or language acquisition device, but instead is part of a communication system shaped by external forces, including information structure, the oral–aural channel, and culture. The second is the claim that if the language lacks recursion then this is a blow to the idea that recursion is what makes human language possible.

Other researchers have taken up this issue. In his 2011 book *The Recursive Mind: The Origins of Human Language, Thought, and Civilization*, New Zealand psychologist Michael Corballis provides a lively discussion of this 'recursion controversy.' Corballis recognizes that recursion is a property of the mind, not of language. The accumulation of evidence in favor of the idea that Pirahã lacks recursion notwithstanding, we still need to consider more carefully how it is that culture could be responsible for the absence of recursion in the grammar of sentences and their parts in Pirahã.

The reasoning to show how this culture–grammar connection works, however, turns out to be relatively simple. We begin with the 'immediacy of experience' principle that I have described in several publications. This principle is vital to the understanding of the interaction between culture and grammar in Pirahã, which requires that evidence be given for assertions. Declarative Pirahã utterances contain only assertions related directly to the moment of speech, either experienced or as witnessed by someone alive during the lifetime of the speaker.

In Pirahã culture everything that is said has to be 'warranted,' that is, there has to be an evidential basis for it. It has to have been witnessed, heard from a third party, or figured out by means of evidence. Not only this, but verbs in Pirahã carry a suffix that signals the source of evidence – **hiai** 'hearsay,' **sibiga** 'deduced or inferred,' and **xáagahá** 'observed.'

The immediacy of experience principle further affects Pirahã's grammar by requiring that the verb, as well as each of the nouns identified in the action of the verb (the subject, object, and indirect object), be warranted by the evidential suffix. Since the verb, subject, and object are expressed by verb and noun phrases, the specific Pirahã link between culture and grammar is that each single set of the verb plus its nouns – the semantic core of each sentence – must be warranted by the verb's evidential suffix. This can be stated more precisely, but it requires us to get a bit technical.

First, instead of talking only about 'nouns' and 'verbs,' we have to use the theoretical term 'nucleus,' which comes from the linguistic theory Role and Reference Grammar. The nucleus is a combination of syntax and semantics and corresponds loosely to what other theories call the 'head' of the phrase, such as a Noun is the 'head' of a noun phrase and a Verb is the head of a verb phrase. Using the term 'nucleus,' we can say that 'Only the nucleus of a phrase specified in the semantic frame of the verb' is warranted by the evidential suffix.

For example, in a noun phrase like 'John's house,' 'house' is the nucleus – the semantic core, what this phrase is about. John is the possessor, a type of modifier of the nucleus, house – the possessor tells us which house we are talking about. On the other hand, in a larger noun phrase such as 'John's brother's house' 'house' and 'brother' are each a nucleus of a containing phrase. 'House' is the nucleus of the phrase 'brother's house' and 'brother' is the nucleus of the phrase 'John's brother.' 'John' is not a nucleus of any phrase. Now since 'brother' is a nucleus that is not the verb's subject, object, indirect object, or the verb itself, it is not warranted by the evidential and thus the phrase violates this culturally imposed grammatical constraint. Linguists refer to the verb, along with its subject, object, and indirect object and accompanying prepositional phrases, as its 'frame.'

In most languages, evidentials are limited to main clauses. In the case of Pirahã this would be because the evidential must warrant each and only the nuclei of the semantic frame of the verb.

If this is correct, then there cannot be phrases within phrases of the same type in Pirahã – there can be no noun phrases inside of noun phrases, or verb phrases within verb phrases, or prepositional phrases within prepositional phrases, because the embedded phrases are not part of the main event described by the main verb and therefore would not be certified by the evidential marker on the verb.

The claim that any language lacks recursion raises a number of issues. Beyond the problems that such a claim brings for universal grammar, it also creates difficulties for some theories of how children learn reasoning skills. One example of the latter comes from influential work by psycholinguists at the University of Massachusetts, Amherst, that develops an idea tracing back to the late nineteenth- and early twentieth-century German philosopher Gottlob Frege. According to this idea, we learn about true and false judgments by the contrast between main and subordinate clauses. In a number of papers the University of Massachusetts research team advocates the view that, without recursion in their sentences, children cannot acquire crucial reasoning abilities, especially the ability to recognize when other people's beliefs are false.

The reasoning goes like this. When we use a sentence to make a simple statement of fact, we can tell whether the sentence is true if it 'fits the world.' If I say, for example, 'John has gray hair' and John's hair is in fact gray, then my sentence fits the world. On the other hand, if John has gray hair and I were to say 'John does not have gray hair,' well, then, my sentence is false because it does not fit the world.

Many psychologists are especially interested in what happens when my belief about John's hair takes the form of a *subordinate* sentence, as in 'Dan says *that John has gray hair.*' Suppose you overheard Dan say this about John. Next you discover that Dan was wrong. Although he said this, John does *not* have gray hair. In that case is 'Dan says that John has gray hair' true or false? It is *true*. Dan might very well be wrong, but if he actually said what you said he said, then the entire sentence is true and the color of John's hair has become irrelevant. Psychologists researching how we talk about false beliefs show convincingly that this flip in determining whether a sentence is true or false can be important in helping children to learn to reason.

When a child learns these structures, she or he can use them to reason about people's thoughts and mental states, as in 'Bill thought that Miranda left so he was depressed,' and so on. Some researchers account for the use of subordinate sentences in reasoning chains created by children by a 'point of view.' When the child understands that what matters is not what Miranda did but what Bill said or thought she did, children learn about other people's points of view. But these researchers claim that such learning takes place *only* when a child learns subordinate sentences. As one of these researchers, Dr Jill DeVilliers, states, 'Before the possession of the appropriate grammatical machinery and key vocabulary (such as the mental state verbs, believe, think, etc.), children may have a range of important understandings of both their own and other people's mental states, but the explicit understanding of the content of false beliefs is not possible.' She claims that without subordinate sentences children simply cannot learn about others having false beliefs or learn to reason about people based on their beliefs. If she is right, then all languages *must* have subordinate sentences or their speakers will be unable to reason well. That is a strong claim. Is it true?

These researchers make reasonable points in grappling with the extremely difficult question of how children come to acquire a theory of mind. But they are wrong to connect this so strongly to syntactic structures. Analysis of the Pirahã language does not support this view. Pirahã children learn about false beliefs without recursion. But this should not surprise us because the same seems to be true for English! I have in mind typical English sentence pairs like the following: 'Miranda left. Or so Bill thought.' 'Miranda left. Well, according to Bill, anyway.' 'Bill was thinking.' 'About what?' 'That Miranda had left.' None of these examples has subordination, yet they convey the same points of view as sentences that do show subordination and recursion.

What emerges from a careful scrutiny of the data, therefore, is that the formal marking of the content of other people's minds via sentence recursion is not necessary, however useful it might be. Just as grammatical markings of all sorts are used in some languages but not others, so it is that no one set of individual marking devices is necessary in all languages for learning to reason about other points of view. Not even recursion.

Our interpretations are based on the meanings of the sentences and the relationship we perceive between them, not on the actual number of sentences or whether one is subordinated to the other *per se*. Sentence subordination does not play a vital role in children learning to reason. Rather, what is crucial is that children acquire the ability to understand discourses. Sentences are part of discourse. The form of the sentences is relevant but not essential.

On the other hand, as our reasoning becomes more complex in certain contexts, such as in placing points of view within other points of view, then recursion in sentences becomes ever more useful. Because English has sentential recursion but Pirahã does not, it turns out that English speakers *can* actually communicate some things that the Pirahãs do not seem to be able to.

English not only has simple sentences that can express a point of view but some very long and complicated ones that can express multiple points of view. Think about sentences like 'John believes the moon is green cheese' and 'John believes that Peter believes that the moon is green cheese.' Those are pretty easy to understand in English. In Pirahã we would have to express the same thought along the lines of 'The moon is green cheese. Or so John believes' and 'The moon is green cheese. Peter believes this. Or so John believes.' But if we go beyond this level of complexity, to something like 'John believes that Peter believes that Bill believes that the president believes that the moon is green cheese,' then without recursion the communicative challenge becomes acute.

Recursion *in the sentence* is thus a tool for expressing longer chains of thoughts within thoughts. Sentential recursion facilitates this kind of complex nesting of points of view. At the same time, the need to express so many thoughts inside other thoughts is relatively rare. I have never seen any attempt to do this in any of the dozens of stories I have collected from the Pirahãs over the years.

Interestingly, even in English we often see alternative strategies for the expression of complex ideas. When the thoughts are part of our cultural knowledge and not unique statements about multiple individuals, we can (and I presume that the Pirahãs could, too) understand them fairly well. Take the following: 'You drink. You drive. You go to jail. Rich people don't go to jail though. That is what some people say.' We could utter this same set of thoughts as a single sentence, using recursion, along the lines of: 'Some people say that if you drink and drive, then you go to jail, unless you are rich.'

We see this type of simple syntax, supplemented by mental operations supplying more complicated interpretations based on cultural knowledge, frequently in English. 'No shoes. No shirt. No entrance.' 'No ticket. No wash.' 'Eat. Drink. Man. Woman.' Among the Pirahãs and in other societies where much is predictable and everyone knows everyone else, such examples are even more common.

Unlike English, which uses both discursive and syntactic strategies, Pirahã relies exclusively on discursive strategies for communicating about other minds. Here is something, for example, that a Pirahã would actually say:

**Kóxoí higáísai. Kohoi hi goó gáísai. Xaogií báaxáí.**
'Koxoi said that. Kohoi said that. The foreign woman is pretty.'

As in English, the interpretations of false belief statements becomes more difficult in Pirahã as the sentence chains grow longer. But, unlike English, Pirahã lacks recursion to help mark the relationships between the different parts. This may be why the successive sentences about beliefs I have in my data do not seem to extend beyond something like the following:

**Kóxoí higáísai. Kohoi hi goó gáísai.**
**Xaogií goó gáísai. Báaxáí, tíi**.

'Koxoi said that. Kohoi said that. Foreign woman said that. I
   am pretty.'

We have learned in this little detour about recursion and false beliefs
that recursion is itself a cognitive tool. Without syntactic recursion the
space for discussing false beliefs and other minds is narrower. We have
also seen that, although Pirahã has structures to express what would in
effect be one to three subordinate sentences in other languages, after
the equivalent of three levels of embedding in English, translations into
Pirahã are nearly impossible. This means that we can never be sure if we
can or have successfully translated some things well between languages.

When the Pirahãs speak Portuguese, as Dr Jeanette Sakel of the University of the West of England has shown in multiple studies, they use
a very small Portuguese vocabulary with their own grammar. In other
words, they are not really speaking Portuguese. Only a very few men
in the culture – and none of the women – speak Portuguese in even
this limited way. Interestingly, there is no recursion even in their Portuguese, according to both my and Sakel's observations. In other words,
in response to the question, 'Can anything at all be translated from any
language to any other language,' the answer seems to be, 'No. Different
languages might have different expressive powers for different kinds of
information.'

The key to understanding how grammar and reasoning fit together,
whether talking about other minds or anything else, is the fit between
a language and its containing culture, society, and situation. What this
all means is that sentence recursion is not crucial to the structures of
human languages and thus, if this is correct, that recursion does not
play the role in natural languages that many linguists have claimed.

Some researchers, however, claim that recursion is crucial and that
even if the rescursion-less analysis of Pirahã is correct, the language is
nothing more than an exception. It is just an oddity. Pirahã is no 'black
swan,' they say. It could be an off-white swan, a slight variation from
the norm that is irrelevant to the theory that all swans are white. Let us
consider this more carefully.

If Pirahã were merely an exception, not a counterexample, then the problem would become even more difficult for nativism, not less so. The problem would be that if we granted that there could be one non-recursive grammar, then we would have to admit that there could be another. But if there were another language that lacked recursion, there could be more.

Most importantly, even if there were just one non-recursive language, then the capacity for constructing recursive sentences could not be what explains why humans have language but other animals do not.

What really matters, however, is not so much the complexity of people's sentence structures, but the complexity of their reasoning. There is no evidence that any communities or individuals *reason* without recursion. There is plenty of evidence, however, that groups like the Pirahãs reason recursively. One bit of evidence comes from their stories.

An early researcher among the Pirahãs, Steve Sheldon, has collected a lot of stories over the years, just as I have. These stories all illustrate recursive reasoning. A representative test from his stories is worth looking at here, especially because he collected these stories in the 1960s and 1970s, long before the analysis of Pirahã became controversial. The following story was first told by a Pirahã woman named Tóóhió and later reported to Sheldon by another Pirahã, Xitaíbígaí. Tóóhió had been to Porto Velho, and had taken a trip to the airport with Sheldon to look at airplanes. While there she saw soldiers and lots of Brazilians. She and her child also saw inside a jet for the first time and the steward gave them some candy.

*Tóóhió's Story about an Airplane*
1. **Xitaíbígai aoaíbá aogagai.**
'Xitaíbígaí there are lots of soldiers.'

2. **Gahisó aíbá ogií hiaigáa.**
'There are lots of big airplanes.'

3. **Toaoaigía xisaitaógi aoigió.**
**abaxaígío hi pio hoagá.**
'Toao [her child] well he all alone with Steve went there.'

4. **Toaoaigía aógió hi ahoáobá boxbóxoi**.
'They gave Toao lots and lots of candy.'

5. **Xitaíbígai gahiaó oogiáaga gáí**.
'Xitaíbígaí there are big airplanes there.'

6. **Gahiaó aíbíbaaxá gahiaó**.
'There are lots of planes.'

7. **Xitaíbígai gahiaó oogiíaga gái gahiaó**.
'Xitaíbígaí the airplanes there are big.'

8. **Sogaga aoaíbaí. Xao abaxaí gaá hoágai**.
'There are lots of soldiers.'

9. **Xao aíbaí ibági ahátaio**.
'Many Brazilians fight.'

10. **Sogaga ao aíbaí**.
'There are lots of soldiers.'

11. **Gahiaó hi gío ábaipí**.
'An airplane landed there.'

12. **Toao aigí ao gió hi ahoáobá bobóxoi**.
'They gave all the candy to Toao.'

13. **Gai hi gahiaó**.
'There are airplanes.'

14. **Koo ao ai ao gió hi ahoáobá bobóxoi**.
'Inside the Brazilians gave him candy.'

15. **Xitaibígai gahiaó aíbá gai**.
'Xitaíbígaí there are lots and lots of planes.'

16. **Gahiaó gahiaó ogiaí aagá. Koobiai gai**.
'There are big white jet airplanes there.' [Literally: 'Airplane, airplane is big. It is white there.']

17. **Gahiaó ao gahiaó akabaí áagá gái**.
'There is an airplane sitting path there.'

18. **Xitaíbígai gahiaó oogiaíi.**
'Xitaíbígaí the airplanes are big.'

This discourse has a recursive structure, showing ideas within ideas, even though none of its individual sentences is recursive. Recursion is found in the divisions of the story and how these fit within one another. The first division is found in lines one and two, in which the teller gives the setting and background of what is to follow. Line three reports the first event and line four reports a second event, which is conceptually subordinated to the first event – getting candy required that Toao first go to the airplane. The second division of the story is made up of lines five through eight. There is a reiteration of the setting. This might seem strange to us, accustomed as we are to different styles for written *v.* spoken discourse. But repetition is a popular stylistic tool in Pirahã. As well as helping them to overcome ambient noise, it is aesthetically pleasing to them.

After the repetition of the settings we see a resumption of the story line, giving more new information. Line nine provides a kind of parenthetical background to line eight. The third division is line ten, which again repeats the setting and resumption of story-telling. Included in this new division are additional events reported in lines eleven and twelve. These lines are followed by a fourth division, very much like the third, with repetition of setting. The fifth division of the little story clarifies things that have preceded it and adds a bit of new information. And the final division of the story is given in line eighteen, where the setting is repeated and then followed by closure of the story.

Lines in the story express ideas that are subordinate to other ideas. The placing of one idea inside of another for the proper interpretation of the story is recursive story-telling. Therefore, recursion must be a property of Pirahãs' thinking – or they could not tell a story like that.

It is worth being reminded that recursion is not some rare commodity found only in human language or reasoning. It is found all over in nature. Pregnancy illustrates recursion: one person inside another; a recursive person.

The ability to think recursively is fundamental to human language because it allows us to speak languages that are non-finite, that have no end. We can talk about anything for as much as we like – the only limit on what we say being our finite brains. The recursive property of Pirahã stories means that their stories have no upper limit. The absence of an upper limit, due to the ability of the Pirahãs to think and structure discourses recursively, means that Pirahã is a non-finite language. However simple its sentence structures, it is a complex language. It is unlikely that any language is finite. All humans have the ability to think recursively, putting one thought inside another to create an infinite and complex array of thought patterns that are denied to other species. However, when this recursion is lacking from the syntax proper, it can affect the intertranslatability of certain ideas between languages, as we saw earlier.

Recursion is just one phenomenon where the subtle reach of culture is seen. There are many other areas of language, however, where we find evidence that languages are shaped by cultures. In an important study, *Deixis, Grammar, and Culture*, linguist Revere Perkins looked at a number of languages and found a strong correlation between cultures and their pronouns. He discovered that the more technologically 'primitive' the culture, the more *complicated* their systems of pronouns. For example, in English we have the pronouns he, she, it, we, they, them, their, I, mine, me, our, ours, hers, his, etc. A small set of less than twenty perhaps. But in some of the languages studied by Perkins, the number of pronouns extends in the hundreds. To give one example of pronouns that do not exist in English, consider a pronoun from the now extinct Tupi-Guarani language of South America, Tupinambá. This language uses a 'fourth person' pronoun. In English, the third person pronouns, he, she, and it, refer to someone we are talking about but who is neither being spoken to directly nor presently speaking. In Tupinambá the fourth person is someone who is neither speaking nor being spoken to and who was not previously taken in consideration in a story or conversation but has suddenly come to prominence. Another kind of pronoun from Tupi-Guarani and other languages is the first person plural inclusive *v.* exclusive contrast. In English, if you are talking to three people and have a lunch planned with two of them only, then at lunch time you will need to say something like, 'Well, let's have lunch. Not you, though. Just those two and me.' In Tupi-Guarani languages you can say either 'We [inclusive] are going to have lunch' or 'We [exclusive] are going to have lunch.' The former means that everyone is invited; the latter that everyone except the person being spoken to is invited. Very convenient. This form of pronoun is mainly found in less technological societies, for whatever reason.

Other areas of grammar that may show close connections to culture come to mind. One is word structure. Three basic strategies for forming words are broadly (and only idealistically, because there are so many exceptions) labeled 'isolating' – as in Chinese; 'agglutinative' – American Indian languages, Turkish; and 'fusional' – as in Portuguese. In a pure isolating system, there is no morphology; that is, there are no endings or beginnings added to words. In an agglutinative system, words add many suffixes or prefixes and each one of these expresses a separate meaning. In a fusional system, individual suffixes have many meanings. For example, in English the suffix -s in 'sings' means that the verb is in the present tense, that the subject is third person, and that the subject is also singular. All in the single sound -s. Some linguists have tried to claim that the existence of such differences across languages prove that language is not a cultural tool. They reason that because, for example, Brazilian culture and Chinese culture are both highly industrialized major players in the world economy, these peoples share cultural values. So the fact that their word structures are so different seems to be an arbitrary fact, unconnected to culture in any way. Even as the Chinese and Portuguese language cultures become more alike, their word structures remain unalike.

But, as I see it, this reasoning is based on a confusion about the nature of culture. Culture is about values and meaning. The research questions of interest here are what and why people evaluate the difference between good *v.* bad and how they extract significance from the world. The answers to these questions distinguish one culture from another.

This is not to say that many values will not be shared across a variety of cultures. Of course they will. All cultures have values similar to 'It is good to treat others as you would have them treat you,' or 'Avoid lying' and 'Take only what you need.'

It should surprise no one that cultures share values. Although the West often criticizes 'Sharia law,' many Sharia values are not unique to Islam and are shared widely among other cultures, including western culture. What distinguishes one culture from another ultimately are two things: the totality of their values *and* the relationships between these values. This latter point is crucial, though often missed.

We can see the effects of relationships, or 'rankings' between values in individuals and entire cultural groups. One group I have had a lot of interaction with during my adult life are *caboclos*, people usually descended from indigenous groups in Brazil who live in semi-industrialized communities along the banks of the Amazon and its tributaries. We might compare these folks with another cultural group, say, factory workers in Ohio. It is fair to say that both groups like big meals with lots of protein. And both value hard work. Likewise, both groups believe that being overweight is bad. But for most *caboclos*, not being overweight has a very high cultural value, since it is associated with sloth and corruption, whereas for many Ohio factory workers being overweight is less of a moral problem and more of a health problem – they do not value being at the right weight all that highly. A ranking like the following seems to describe the two groups. I borrow from linguistics the symbol '>,' which means 'outranks:' *caboclo*: hard work > overweight is bad > good meal; Ohio factory worker: hard work > good meal > overweight is bad. At least in principle, then, two societies can have the same values but show marked superficial differences that follow from the relative rankings between these values. Broad comparisons between societies and cultures fail to discover this type of value ranking. Rankings and the values that they order from top to bottom cannot be discovered through questionnaires but require deep, long-term observation and participation in the cultures, for instance, comparing what people say with what they do. As with any theoretical tool, the usefulness of this method can be tested by its efficacy in describing our findings and by the predictions it makes about people's behavior, appearance, and so on in different groups.

Only with this complex conception of culture does it make sense to claim that language is an instrument for solving the general problem of communication in conformity with the values and the rankings between values of specific cultural groups.

## Chapter Twelve

# WELCOME TO THE FREAK SHOW

*'Familiarity breeds contempt, while rarity wins admiration.'*

Apuleius

Long live diversity. We need outliers, free thinkers, freaks, non-conformists, mutations, and exceptions to constantly remind us that there are alternative ways of doing, living, and being in the world. A source of linguistic and cultural diversity is found in the variety of languages spoken in the world today. Another source is the range of peculiarities to be found in each individual language. The diversity of languages spoken on planet earth is one of the greatest survival tools that human beings have as a species. This importance stems from the fact that each language is a cognitive tool for its speakers and comes to encode their solutions to the environmental and other problems they face as a culture. Collectively, therefore, languages express the collective wisdom of our species in ways that are never fully translatable in their entirety from one language to another. In the course of human history, however, many peoples have changed tools, that is, they have switched languages. When all the speakers of a language switch to another language, then the former language has died. This is how linguists describe a language that is no longer spoken, whether or not it is written – it is a dead language.

If we think of language death as a shift from one cognitive tool to another, though, does it really matter if people abandon one implement for another which, for whatever reason, they have decided works better for them? Not in the abstract, even though it could be the case that this change in tools was forced upon them by, say, a conquering army. But will it really matter in a hundred years that such-and-such a language has died?

Yes, as a matter of fact it will. It is a matter of the gravest concern. There are many reasons to be alarmed at language loss. The main reason is that all languages are tools not merely for their community of speakers but for the entire human race. There are about 6,800 mutually unintelligible languages spoken in the world today. Many languages spoken in the past have ceased to exist and many more languages will come into being in the future. Since the arrival of *Homo sapiens*, new languages have been constantly emerging as others have vanished. This is why many linguists say that the total number of actual languages spoken in the world at a given moment of human history is but a small fragment of the much larger number of *possible* human languages.

Given this continuous cycle of death and rebirth, it might seem that the death of one language is not a particularly serious event. Who cares if some little island language has died off? How does this fundamentally affect humanity one way or the other? The answer is ultimately that each loss is a terrible human and scientific tragedy. A language is a repository of the riches of highly specialized cultural experiences. When a language is lost, *all* of us lose the knowledge contained in that language's words and grammar – knowledge that can never be recovered if the language has not been studied or recorded. Not all of this knowledge is of immediate practical benefit, of course – it will not all lead to new medicines, technologies, or means of emotional development. But it is all vital in providing us with different ways of thinking about life, of approaching day-to-day existence on planet earth.

In my thirty-plus years of field research on languages of the Brazilian Amazon, I have had the privilege of living for more than eight years in villages of the Pirahã, with additional months among the Banawá, Paumari, and other groups. The Banawás are members of a select group of Amazonian Indians that make *curaré*, a fast-acting and deadly strychnine-based poison used on blowgun darts and arrows. The ability to make this poison is the result of centuries of knowledge-gathering and experimentation which is encoded in the Banawás' language in its use of terms for plants and procedures. However, that is in danger of being lost, as the last eighty or so remaining speakers of the language gradually switch to Portuguese.

The Pirahãs speak a language unrelated to any other living language.

That alone makes it a rare jewel to be studied. It is famous among linguists around the world for many characteristics that have shaped our knowledge of language in interesting ways. But the Pirahãs have begun to ask me each time I visit why they are so few (they say so 'small') and why there are so many 'crooked people' in the world. (The Pirahãs view all non-Pirahã people as 'crooked,' that is, bent and not working properly.) They only speak Pirahã (aside from the bit of Portuguese spoken by some of the men) and they reject outside languages. They sum this up as 'Pirahãs speak Pirahã.' For a Pirahã to lose their language, therefore, would be to lose their identity, their perceived means of survival in this hostile world.

There are many groups like the Pirahãs and Banawás, whose languages are threatened by turmoil, loss of identity, and marginalization. Two others are the Oro Wins and the Suruwahás. Decades ago, the Oro Win (odo-WEEN) people were enslaved like the ancient Hebrews, but by Brazilian rubber traders rather than Egyptians. The few dozen survivors of these years of suffering just managed to escape their 'patrons' in the last half of the twentieth century, going to live with the Wari', speakers of a related language in the same linguistic family (Chapakura) near the Mamoré river of western Brazil.

In 1995, I arranged to work with three of the last five speakers of the Oro Win language. I was able to draw the attention of the linguistics community to the fact that Oro Win was a distinct language, not merely a dialect of Wari'. I also noticed that the Oro Win had forgotten much of their language, since their circumstances over the years had forced them to rely on a mixture of Portuguese and Wari', rather than Oro Win, in order to survive. Yet they speak neither of these foreign languages well. The five remaining speakers of Oro Win now find that not only are they unable fully to recall their own language, but they cannot speak any other language as native speakers. They have lost their history and their community. Although they are planting their fields, intermarrying, and getting on with their lives, within the next decade or so the Oro Win language and all vestiges of the Oro Win as a distinct people will be lost to them and to us. And what will we have lost? We will have lost their knowledge of the jungle, a philosophy of life, paths to happiness. We will have lost their literature, art, and scientific knowledge. We will also have lost knowledge of language structures that we will never find again. We will never be able to Google a lost unwritten language. When they go, they are gone for ever.

There are other reasons that languages are lost, and such losses diminish us all. As a final story of the loss of a human language, I want to turn to another small Amazonian tribe, the Suruwahás (soo-doo-wa-HA). This group speaks a language of the Arawá family. Their current circumstances illustrate the worst kind of tragedy that can strike when a language and culture are threatened. The Suruwahás, like the Banawás, to which they are linguistically and culturally related, display a spectacular knowledge of their environment shown, among other things, by their ability to make lethal poisons which they use mainly on their arrows and blowgun darts.

However, in the years following their first contact with the outside world, in the early 1980s, they have begun to commit suicide, drinking their own *curaré*. Out of a population of only a couple of hundred, in one recent year eight adults and teenagers committed suicide on the same day. No one fully understands why these people are beginning to kill themselves. But the answer seems to be related to their sense of fragility and smallness as a people, to the idea that their language, culture, and values cannot compete with those from the outside. It is as though they take the death of their language to be the death of their people.

When they lose their will to live, when the languages of these endangered peoples pass from history without a detailed record, the whole of humanity loses an element of the tapestry from which our history is torn. Our species is less smart, less robust, and less healthy, and more threatened after the death of *any* language or culture.

For many people, like these Amazonian groups, the loss of language brings loss of identity and sense of community and loss of traditional spirituality. To save languages like Banawá, Pirahã, Suruwahá, Oro Win, and hundreds of other endangered languages around the world will require a massive effort by linguists, anthropologists, and other interested individuals. We need, as a minimum, to identify languages which are at risk and then we must learn enough about each of them to produce a dictionary, a grammar, and a written form of them, to train native speakers of these languages as teachers and linguists, and to secure government support for protecting and respecting these languages and their speakers. A daunting task. But it is vital.

At the same time, new languages do still come into existence. There are places in which the right conditions arise for new languages to be born. Observing these incipient languages is important because of what each tells us about the utilitarian nature of language. One example I am familiar with comes, again, from the Maici river in Amazonas, Brazil.

When Brazilian traders boat up the Maici to Pirahã villages, the traders say things to them like:

*Cuyã porasi muito. Papai coxixi pouco.*
kooYAN pudaSEE MUIto. paPAI kosheeSHEE poKU
'The women danced a lot. Dad slept little.'*

To which the Pirahãs might respond:

*Pirahã kabahá casá. Coxixi muito. Doente muito. Korumi fome.*
Pirahã kabaHA kaSA. kosheeSHEE MUIto. doENchee MUIto.
    koruMEE FOmee.
'Pirahãs not hunt. Sleep much. Sick much. Child hungry.'

The entertaining aspect of this exchange for me is that the Brazilians believe that they are speaking Pirahã (and will swear to anyone that they speak Pirahã if asked) and the Pirahãs believe that they are speaking Portuguese. But neither of them is speaking either Portuguese or Pirahã. What they are in fact doing is using a pidgin, a mixed, baby-talk kind of invented language. In the exchange above, the word *Cuyã*, for example, is a Tupi-Guarani word meaning 'woman' – it is neither Pirahã nor Portuguese – and *porasi* comes from the Tupi-Guarani word *porasey* 'to dance.' *Coxixi*, on the other hand, comes from the Portuguese word *coxilar* meaning 'to nap' and *kabahá* is possibly a Tupi-Guarani word that has come to mean, in this context, 'do not have' or 'is lacking.' Or it might be a corruption of the Portuguese verb, *acabar* 'to run out of or to finish' something.

---

* 'Dad' is a term used by Pirahãs and traders for Brazilian river traders.

Thus this exchange in the pidgin language used between Brazilian riverboat traders and Pirahã men has no Pirahã in it at all. It is a mix of Tupi-Guarani and Portuguese. The Tupi-Guarani words in it come from a nearly extinct trade language called Nheengatu 'Good Tongue,' which was the trade language of the Amazon for generations and which was at one time as widely spoken as a native language in Brazil as Portuguese. The Portuguese found in these exchanges is not regular Portuguese, not even like the Portuguese normally spoken by the river traders. For one thing, the verbs in the exchange above are not inflected, appearing almost exclusively in the form of infinitive verbs, like English 'to run,' 'to buy,' and so on.

Pidgins are common around the world, illustrating the point that language tools will be invented as they are needed. Some languages today that started out as pidgins have now become full-blown languages. Tok Pisin 'Talk Pidgin' in Papua New Guinea is one such. Pidgins are *invented*; they are communication tools distinguished functionally from full languages because they serve in a much narrower range of contexts, usually trade. They are found around the world, always in circumstances similar to that of the Pirahãs and Brazilian river traders.

These crude tools are invented because people need to communicate. Therefore, they will find ways to communicate when they need to. Not all languages invented are pidgins, however. Computer languages are invented languages. Formal logic is an invented language. Children occasionally invent languages for games, such as 'pig Latin' ('*U-day o-nay ig-pay atin-lay?*' 'Do you know pig Latin?').

Some full languages are also claimed to be 'invented by' or 'suddenly appear from' children who are not native speakers of any language. Creole languages exemplify these. And some languages are invented by individuals or groups for utopian or functional goals. A beautiful study of invented languages is provided in Arika Okrent's 2009 book *In the Land of Invented Languages*, in which she discusses languages like Esperanto, Klingon, and many others. What is especially interesting about all invented languages is that they are directly or indirectly based on existing human languages and that they only become full languages – on a par with English, French, Korean, and so on – when they become a *native* language, that is, when they are spoken as the first language by some speakers and are used as the normal medium for daily conversation. Tok Pisin of New Guinea is such a language. To my mind, invented languages are excellent evidence for the theory that language is a cultural tool. But not all agree with me.

The best-known proponent of the view that creoles (traditionally defined as pidgin languages that have become someone's native language) prove that languages are structured based on innate, language-specific principles is Derek Bickerton. He has written extensively about creole languages and universal grammar and his leading ideas have come to be known collectively as the 'language bioprogram theory.'

With these languages, a process called creolization takes place when children adopt a specialized, but highly unstructured and largely deficient pidgin language as their native or first language. This happens when the children are, for one reason or another, unable or unwilling to learn their parents' language, learning instead only the pidgin language that a previous generation developed merely for trade purposes. The process of creolization is relevant to our understanding of the nature of language because, according to the bioprogram theory, children's biologically programmed, universal grammar transforms the crippled pidgin into a robust, real, complete human language, a creole.

This theory uses the typical range of nativist arguments to make its case, the main point being that a creole language has all the characteristics of any 'whole' language and that it is far richer than a pidgin. The supposition is that these features could not have been learned. The claim is that learning is unlikely because the creole contains features that are not present in any of the languages that the child has been exposed to. Since the features that distinguish the creole from the pidgin could not have been learned, so the story goes, they must be innate, part of the children's bioprogram. I do not agree.

Of course, no one argues with the idea that if a language is the only language of a group of people, then it will need to fulfill all the functions required of it to work. There will be enormous pressure on children to make their language whole by invention or by borrowing from other languages. In fact, the possibility of borrowing is a potentially serious problem for the bioprogram theory. Unless it can rule out both the possibility that creole features were taken from existing languages and also the possibility that its features are straightforward solutions to common problems of communication, then there is no reason to suppose that creolization is simply the unfolding of a bioprogram. To date no proponent of the bioprogram theory has ruled out these alternatives.

In fact, however, the case for the bioprogram theory has more serious problems. Two of the world's leading experts on language contact (the phenomenon of linguistically consequential interaction between speakers of distinct languages) have discussed the sudden appearance of creoles in detail. These authors refer to the process as 'abrupt creolization.' Their analysis is at odds with the bioprogram theory. These researchers, Dr Sarah Grey Thomason of the University of Michigan and Dr Terrence Kaufman of the University of Pittsburgh, have found no evidence for an innate grammar in abrupt creolization. In fact my interpretation of their work is that it supports instead the hypothesis that languages are cultural tools, developed through a combination of invention and building on past knowledge. According to these authors, creoles arise for a variety of reasons, within a range of distinct linguistic and sociological environments.

Pitcairnese is one of the creole examples they examine in making their points. Pitcairnese was first spoken by the children of the mutineers from HMS *Bounty* in the early nineteenth century. These children's parents came from nine English-speaking *Bounty* crewmen and sixteen Tahitian speakers. Although only these two languages were spoken, the children came up with their own language, Pitcairnese, named after the island of Pitcairn, where they were born. There was no well-developed pidgin on Pitcairn island, yet these children 'invented' a brand-new creole language for their native tongue.

Pitcairnese demonstrates that creoles can arise abruptly in order to serve as communication tools in specific environments. The study of Pitcairnese also shows that the bioprogram theory rests on problematic premises. The bioprogram theory assumes that all of the creoles that are created by children learning them as a first language share a set of grammatical features, presumably because they are innate. It is further assumed that these shared features cannot be explained merely as the borrowing of features of the languages spoken by the children's parents or other persons of significance in the children's environment.

Thomason and Kaufman show that Pitcairnese displays features of both English and Tahitian. Clearly the child inventors of Pitcairnese knew a good deal about other languages. This is a problem because, according to the bioprogram theory, the core features of creoles cannot have been learned.

In the end, there just is *no* grammatical feature that is found in all creole languages. We do not observe the degree of worldwide homogeneity in these languages that the bioprogram theory predicts. Also, there is evidence that all creoles share influences from the variety of languages in the history and environment in which each arose – the parents' languages, pidgin languages that speakers of creoles know independently, and so on. This research offers compelling reasons to believe that influence from other languages on creolization is always felt.

Finally, although they do not put it in these terms, they make a case that creole languages involve not only the influence of other languages but that many of their features are the simplest solutions to the array of problems any incipient communication tool would face as it tried to fill the role of a native language.

The evidence leads to the conclusion that the bioprogram theory is based on an unnecessary appeal to nativist principles. Humans are good at solving problems and inventing tools. Their language tools arise in the context of the gamut of human knowledge, experience, and social constraints.

The other source of linguistic diversity, freaks hiding in individual languages, is found in rarities of form and meaning discovered within, rather than across, the languages of the world. Not the gross total diversity of vocabulary and grammar but specific jewels found only after enormous analytical effort. Intellectuals have long recognized the importance of diversity. The pedigree of the philosophical understanding and appreciation for the importance of diversity is at least as old as the colonization of America.

On February 5, 1631 Roger Williams and his wife Mary landed in Boston from London. This young, new minister was destined to change the course of American history and American philosophy. Williams began almost immediately upon his arrival to irritate the church establishment. Among the irritants was his preaching that the church had no right to impose its theology via civil law. The modern-day equivalent of the situation in which he found himself might be a Muslim imam telling other imams it was wrong to impose Sharia upon a non-Muslim population. Williams taught that every individual had the right to live according to their own beliefs and that the Puritans should not try to enforce their own morality on others. He was the first person in America to argue for a 'wall of separation' between church and state. In this and another way we are about to see, he influenced Thomas Jefferson and, through Jefferson, other founding fathers of the United States, preventing the new American colonies from living in the theocracy that the Puritan leadership so desired to impose.

Roger Williams is a father of American tolerance. I was convinced of this by an intriguing 2002 book, *Native Pragmatism: Rethinking the Roots of American Philosophy* by Scott Pratt, a philosopher at the University of Oregon. Diversity is not incompatible with unity, of course. At least in theory, America believes that *e pluribus unum* – out of many beliefs, cultures, histories, and religions can emerge one new country, unified yet diverse.

But among Williams's most radical applications of tolerance were his belief and teaching that the native inhabitants of America had independent rights and that no European government or settler had the moral authority to dispossess the indigenous population of the new world from its lands. He accused King James of lying about his rights to their lands – no European monarch, nor any American settler, had the right, Williams claimed, to issue or use legal charters as a pretext for the confiscation of Indian lands.

What happens to dissidents in a culture that highly values uniformity and central authority? They get into legal trouble. In 1636 the law came for Roger. They meant to punish him severely for his sedition – his demand for tolerance, diversity, and indigenous land rights. But before the sheriffs could lay their hands on him, Williams escaped, into a cold Massachusetts January, making his way on foot more than one hundred miles to Naragansett Bay, where he was received by his friends, the Wampanoag.

The Wampanoag accepted Williams warmly and helped him to develop his philosophy of tolerance even further, according to Pratt, by exposing him to their own concept of **Mohowaúsuch**, a sort of 'canni-bal' or 'monster.'* The Wampanoag word **Mohowaúsuch** was the name of a specific creature who was not like humans, but was tolerated, and not uniformly feared, by the people because it represented to them one extreme of human life. The Wampanoag valued diversity.

This tolerance and valuation of diversity are not unique to the Wampanoag among American Indians however. It is seen many times. The Pirahãs have a similar, every bit as remarkable, tradition of tolerance for monsters, freaks, and the rest of the gamut of human experience. Like the Wampanoags, the Pirahãs believe that there exist real, non-fictional humanoid entities. The Pirahãs call these the **Kaoáíbógí** 'Fast Mouths.' The Kaoáíbógís look like humans, though they often bear some sort of deformity. They also always speak in a falsetto. The Pirahãs believe that these creatures can be either mean or kind and that they will have sex with Pirahã women at the first opportunity. In spite of the Kaoáíbógís' moral and physical strangeness, though, the Pirahãs toler-ate them and appreciate them as entertaining, diverse life forms, just part of nature's bounty.†

According to the Pirahãs, they see Kaoáíbógís almost daily. These creatures talk funny and look like Pirahãs, but they can be mischie-vous. They eat different things – fruits and animals that the Pirahãs themselves do not eat, such as snakes and lizards. But the Pirahãs call to them and leave things out for them. I once referred to these enti-ties as 'spirits.' In fact, there is no word to translate them into English. They are tolerated, occasionally fearful, occasionally frightening living creatures. A Pirahã will refer to a shadow out of the corner of their eye as they walk through the jungle as a Kaoáíbógí. Many things that they cannot explain they attribute to the work of Kaoáíbógís. But the princi-pal manifestation of Kaoáíbógís to the Pirahãs are their fellow villagers,

---

* This term was the pejorative term used by the Wampanoags for another famous indigenous group, the Mohawks, referring to them as 'the monsters' or 'the cannibals' – not a term that the so-called Mohawks liked. This people's autodenomination was **Kaniengehaga** 'People of the Place of the Flint.'
† These are not fictional. The Pirahãs see them. So do I. When I remark that this or that spirit that has come to the village at night, say, to dance, looks like a Pirahã man I know, they respond simply that 'they look like Pirahãs alright.'

who will speak in a falsetto voice and come into the village naked at night claiming to be a Kaoáíbógí.

Besides the Kaoáíbógís, the Pirahãs' tolerance can be seen in various other ways. The way that they treat outsiders exemplifies acceptance and enjoyment of diversity. They love to ask me what I am doing, why I am doing it, whether other Americans do it, and so on. At times their questions make me think I am being interviewed by a team of anthropologists. They never criticize, they never condemn, they never tell me that there is a better way to do things. They just enjoy watching me and other visitors be different.

One hears stories of this or that people who are offended if outsiders will not eat what they offer them or will not participate in their rituals or who live according to different customs even during their visit to the people's community. Not the Pirahãs, though. They offer food, but are not surprised if I turn it down. They do not want to eat everything that I eat either. Women might offer to have sex with me, but are not offended if I say no. They do not mind if I sleep in a bed or wear shoes or build an outhouse. They just say things like 'This is how other people do these things. They are not Pirahãs.'

According to Pratt, this American Indian value of tolerating monsters and diversity became a foundational concept in the origination of the single most important contribution of American thinking to world philosophy – pragmatism. The Wampanoag influenced Williams, who influenced Thomas Jefferson, who indirectly influenced William James and other founders of the school of American pragmatism.

Pragmatism is based partially on the idea that there is no one intellectual 'ring to bind them all.' Pragmatism denies that there is 'Truth,' with a capital 'T.' It denies that science, philosophy, religion, or any other form of inquiry should seek truth as the holy grail of human study. Rather pragmatism argues that while truth may exist some place in the universe, we mere evolved primates must be content with finding ideas and theories that *work* for us, that serve as tools for our different tasks.

There is an important connection between Roger Williams, pragmatism, tolerance, and theorizing about human language. But I want to say some more first about tolerance and pragmatism, because these play such an important role in my own view that language is a tool, as well as helping us to understand why some theories are intolerant.

American Indian tolerance stands in stark contrast to the European and North American valorization of conformity. Perhaps all creatures, including humans, prefer most those beings that look like themselves. That could have been an effective survival tool in the early days of *Homo sapiens*, as it is for other creatures (cats like dogs less than other cats). We like homogeneity. In European culture and its descendants, this often extends to points of view and philosophies. We enjoy ideas that sound more like our own. Perhaps most of us are, deep down, like John Wayne when he said, 'If everything isn't black and white, I say "Why the hell not?"'

The love of homogeneity transfers to some intellectual endeavors as well. Scientists can be intolerant of diverse ideas, even though many will deny this. Yet is should not surprise us to learn that science is a culture with its own ranked values. It is not a purely intellectual endeavor. These points are made effectively by philosopher of science Thomas Kuhn in his famous 1962 book *The Structure of Scientific Revolutions*. The popular English expression 'I have a new paradigm' comes directly from Kuhn's book.

Perhaps the most important value shared by all scientists is 'Be objective.' Others include 'Work hard' and 'Let the chips fall where they may.' But not all values are equally positive. All cultures have values that other cultures might disapprove of or at least rank very low. One interesting and highly ranked value of science is that knowledge and theories should be constructed homeopathically, bit by little bit, upon the acknowledged work of your predecessors. Other values, less highly ranked, but still important include things like 'Do not contradict your professors' and 'Accept your station.' By examining the full range and ranking of the values of scientists, we can better understand not only the outputs of their enterprise, but also how they arrive at those outputs.

To illustrate the value 'Do not contradict your professors,' imagine that a professor became famous for proving that all birds can fly. You learn this and head off one day for your own field research in Antarctica. While there you discover penguins. They cannot fly. Now how do you report your discovery? If you rigidly follow the unspoken mores of scientific society, you will not say that you have found a counter-example proving that your professor is wrong. Rather, you are likely to claim that you have discovered an anomaly and that this discovery could turn out ultimately to support your professor's view of avian flight. That way you will be invited to the best venues to discuss your curious and entertaining findings. But call penguins a counterexample and you could run into problems. You could even find that you have become anathema to your former professor and colleagues. You might have to take a metaphorical one hundred-mile hike in the dead of winter to find life in another society. Exception or counterexample? That is the question. But your answer will have social significance.

Nevertheless, the discovery of empirical outliers (and even their attendant controversies), however recognized, has value for all scientists in establishing the boundaries of human knowledge, occasionally even revealing that there are some things our theories can never, even in principle, account for.

In 1996 the late UCLA phonetician Peter Ladefoged and I published in the flagship journal of linguistics, *Language*, an article entitled 'The Problem of Phonetic Rarities.' It was about some rare sounds I had previously described and their consequences for theories of language. Two of the sounds are from Pirahã: a voiced bilabial trill, written phonetically as [B]; and a unique sound that does not occur in any other known language in the world, a voiced, linguolabial lateral double-flap, [ǃ].

The bilabial trill is found in a couple of other languages in the world besides Pirahã, one being the Mexican language Isthmus Zapotec. But this sound is frequently made by little children of all languages on their way to acquiring the adult pronunciation of their language. It is also used in play – I have seen some American children produce the sound to imitate a motor. It is produced by holding the tongue motionless and expelling air out of the mouth in such a way as to get both one's upper and lower lips flapping like flags in the wind, while the vocal cords are vibrating.

As the linguolabial is unique, a symbol had to be invented for it by some missionaries who preceded me to the Pirahãs. This sound is a voiced, labio-laminal alveolar double flap. That long name means that it is produced by first setting the vocal cords in vibration, as air passes through them from the lungs out the mouth. Next, the tongue tip 'flaps' across the upper hard part of the roof of the mouth on the small (aleveolar) ridge behind the teeth. Then it passes between the teeth. On its way the lower blade of the tongue must hit the bottom lip. The visual effect of the tongue coming out of the mouth like this is something like sticking out your tongue at someone, though pointed downward.

I had heard of this sound from Steve Sheldon when I first entered the Pirahã territory. I was keen to observe them produce it, so I drew up a list of words and asked the Pirahãs if they could help me learn these in Pirahã. I would purposely mispronounce them, hoping and expecting that the Pirahãs would correct my pronunciation and give me my first 'glimpse' of this sound that is found in no other language of the world. When they repeated after me or corrected me, however, they did not use the linguolabial but a simple 'g' sound. I also knew that the unusual linguolabial sound could be replaced in some contexts by the 'g', but they were *only* using the 'g.' What was going on?

One night after I had been living in the village for about six months, my main teacher, Kohoi, came into my hut and, in front of about a dozen Pirahãs, said, 'You have asked how to say the word "milk." Now you live here like a Pirahã. So you should say this sound.' And he proceeded to make the linguolabial sound. Many times. It was a wonderful evening. I felt as though I had found gold.

It turned out, as I discussed this sound and collected more examples over the next several days, that the Pirahãs were sensitive to the fact that Brazilians did not have this sound and that river traders made fun of them when they used it, claiming when they did that the Pirahãs spoke like chickens. So it had become a sign of cultural belonging, no longer merely just another phoneme. A cultural object, the linguolabial, had taken on a secondary cultural value.

Now here is where the linguolabial gets really interesting, though. Is this freak an exception or a counterexample to theories of human speech sounds? None of them predict its, after all. Or is it something else altogether? The quandary is this. If theories can predict the existence of the sound, why is this the only instance of it in the world? In other words, can a theory predict both the existence of a sound *and* its rarity? That is, can a theory tell us that something should exist but that we will not be able to observe it frequently in nature. That can be harder to do than you might think. For these sounds, Ladefoged and I concluded that no theory can. Well, then, should we tweak the theory a bit to get it to include this sound, even though it is unique? Or should we say that it is so rare that it should not fit into any theory?

Of course, we can always make something fit our theory by stipulation, that is, by listing it as an exception and saying something like 'We predict that all swans are white. Except for this black one here.' That is somewhat like the way in which in Greek mythology Procrustes fitted guests to his bed – by stretching their limbs when they were too short or by chopping off their limbs when they were too tall. But if we take this tack, we learn nothing. We cannot just ignore a counterexample. If we do not include it in the theory, we predict that it does not exist, which is false. If we do, we have to make an ugly stipulation, which makes the theory look bad. What to do?

There are two options. The first is to treat the anomaly as counterexample. This option means that our current theory is wrong. The other option is to treat the freak as an exception that nobody's theory should worry about. In effect, this option can be called 'theory tolerance.' It means that we do not abandon theories just because they are wrong here and there. If a theory works well enough then perhaps we should not expect it to handle all outliers. This is the Wampanoag/Pirahã answer – conceptual cacophony or theory tolerance. It is also the pragmatist analysis – it works, but there is no true theory possible. We are not right if we do not include it. But we can never be right enough to explain everything anyway – so let's just live with the freaky thing. Whatever we decide, such examples do warn us not to take our theories too seriously.

Unique sounds like the Pirahã linguolabial are like gravy on a scholar's chin. We do not want them there. We would like to be neat and clean and have our theories be the same. But you can't always get what you want, as Mick Jagger reminds us. We cannot explain everything.

Maybe if there were a Pirahã phonologist, she would develop a theory that could both predict the existence of its rare sounds and explain why they are so rare. But then she might as easily fail to predict the rare sounds of English, such as our 'th' sound, as in 'the,' 'though,' 'then,' and so on, or the different but related sound of 'throw,' 'thought,' and 'thick.' Or maybe she would have a problem explaining the German umlaut vowels.

In the paper that Ladefoged and I co-authored we concluded that such sounds are clearly 'sounds that [have] to be described ... We must not overlook [them].' We went on to say that such sounds are not 'describable in terms of the set of features required for interpreting general linguistic processes in a large number of languages.' To put this another way, no theory can account for everything. No theory can tell the whole truth about human speech sounds. And the same holds for syntax and other components of language. There are types of phrases, sentences, and other constructions found in only one or two languages of the world. Their implications for linguistic theory are just the same as for speech sounds – they are unassimilable outliers, dirt smudges on the theoretician's new white lab coat.*

The Amazonian language Wari' is chock-full of rarities. In fact, researchers at the Max Planck Institute for Evolutionary Anthropology declared at a conference on language rarities that Wari' have the greatest number of rarities known in a single language. They based their claim on the grammar that I co-authored with New Tribes missionary Barbara Kern. Like Pirahã, Wari' has a highly unusual sound, [tB], one not found in any other known language other than the closely related Oro Win. This and the many other rarities of Wari' are largely resistant to theorizing and are unlikely ever to be brought under the scope of any linguistic theory.

What rarities do, or should do, is to highlight the individuality, the personality, of each language, since all languages turn out to have rare jewels found in no other language. Edward Sapir described the 'genius' of a language as the way in which that language holds together. No two languages have identical 'geniuses.' Each language is therefore, to Sapir, unique. And not just through the odd sound here and there. Like

---

* Many researchers make a point of searching out such 'smudges' and outliers. But we can never explain all anomalies nor can our lab coats ever be completely clean.

Pirahã's lack of recursion, the genius of a language is what makes its grammar or discourse different, what makes it a whole system.

Therefore, if a linguist describes a language in such a way that, aside from its individual words, it sounds roughly like any other language, then in a way she has missed its genius and, in my opinion, has produced a poor, superficial description of that language that does science a disservice. We cannot learn about the human condition by homogenizing it.

Yet homogenization is *so* tempting. Generalizations and theoretical simplifications are fundamental to our ability to survive in and understand the world. Never doubt their usefulness. But they are neither more important nor logically prior to the basic work of carefully describing the uniqueness of each language, what holds it together in Sapir's sense. And what holds all languages together is the connection between each language's grammar and its culture.

The problem of linguistic rarities for many linguists is that they are outlaws – they refuse to follow theoretical laws. Although some linguists do not mind rarities and find them interesting in their own right, others see them as a bother from which few, if any, lessons on linguistic theory need be drawn. The contrast between Sapir's interest in accounting for the genius of each language versus the drive to discover generalizations among contemporary theoreticians leads us to a philosophical divide that is found in linguistics, anthropology, and many other sciences, the distinction between the 'nomothetic' and the 'idiographic' subcultures of rational inquiry, as found in the writings of one of the founders of American anthropology, Franz Boas. The former are interested in finding quantitative laws, the latter in providing fuller, more nuanced qualitative accounts.

As Boas understood the issues, 'Nature is various, temporal, and individual … It is not justified to conclude from the defects of the available descriptions, which do not reveal a unity of culture, that the whole culture must be a compact unity, that contradictions within a culture are impossible, and that all features must be part of a system … Have we not reason to expect that here [in primitive cultures] as well as in more complicated cultures, sex, generation, age, individuality, and social organization will give rise to the most manifold contradictions?'

Like his pragmatist colleague at Columbia University, John Dewey, Boas asserted that the object of study in science should be the individual elements of complex phenomena. Wholes are abstractions, to be taken with a grain of salt. In his 1887 paper 'The Study of Geography,' Boas considers his two types of scientists in more detail. He laments those who consider individual phenomena worthy of study, not for their own sake but only as 'proof or refutation of their laws, systems, and hypotheses. Losing sight of the single facts [this type of scientist sees] only the beautiful order of the world.' On the other hand, he celebrates the scientist who studies a particular phenomenon because it 'may occupy a high or a low rank in the system of physical sciences ... [and] lovingly tries to penetrate into its secrets until every feature is plain and clear. This occupation with the object of his affection affords him a delight not inferior to that which the physicist enjoys in his systematical arrangement of the world.'

Boas refers to the former as nomotheticists, those who look for laws. This group understands the study of language to be primarily a quest for general principles about language. They want to systematize the 'blooming, buzzing confusion' of the world, as William James put it, that we encounter via our senses. They seek to simplify it.

Boas refers to the second major grouping of scientists as idiographicists: those who write about what makes their objects of study unique. These folks do not look for simplification through law-like generalizations, but for accuracy of narrative – telling the best and most complete story about the facts, so as to bring out the distinctiveness, the peculiarities, and the boundaries of each language. The idiographicist movement founded by Boas, propagated through his students Edward Sapir, Margaret Mead, Ruth Benedict, and others, led to an extremely fruitful period in American linguistics.

The nomotheticists and the idiographicists need not be in conflict, however. In principle, there is nothing contradictory about combining the two perspectives. Yet in practice they are very different – your personality will likely better fit one or the other. But if you are one or the other, you choose to spend your time and thoughts either looking for laws or looking for 'genius' – it is good to look for both.

Is there any explanation for anomalies that escape the boundaries of our theories? Linguistic rarities are marks of cultural distinction. We do not need to say that they were produced in their current form to satisfy some cultural need. They may have been. But probably for most, at least many, rarities we will never know. But what can be said is that such rarities mark a language like beauty marks on the face of a model – they are the seal of uniqueness of a language, a culture, or indeed a human being.

At the same time, we have seen that there are uses of icons in sounds and syntax that arise from the connection of language to culture, such as the Pirahã women's smaller articulatory space, idiomatic expressions like 'kick the bucket,' and the meanings of words like 'haggis' that are so clearly culture-specific. Thus language rarities bring us face to face with the mirror and the reality of the shortcomings of our deepest efforts to generalize. These are our monsters of tolerance and so, like the Wampanoag, we must learn to accept and enjoy the freaks we encounter in the behaviors and utterances of human societies.

# GRAMMARS OF HAPPINESS

*'There is but one truly serious philosophical problem, and that is suicide. Judging whether life is or is not worth living amounts to answering the fundamental question of philosophy.'*

Albert Camus, 'The Myth of Sispyhus' (1942)

The twentieth-century French philosopher and writer Albert Camus asked why, when faced with the shortness and apparent pointlessness of our lives, we do not all kill ourselves. Or, from the opposite perspective, if our lives are precious and our only experience is of happiness, why does *anyone* commit suicide? Camus took up the question of life, happiness, and the specter of death in his 1942 essay 'The Myth of Sisyphus.'

Sisyphus is the character of Greek mythology with whom I most sympathized when I discovered his story in my pre-adolescent years. He was condemned by the gods to roll a boulder up a hill, only to watch it roll down at night and begin again the next morning. And so went every day of his eternity. What could be a worse hell? I thought to myself the first time I read this story. That was before I worked for the US Postal Service and saw the mailbags that I had emptied the night before, full again each new morning.

Whenever I visit an Amazonian tribal society with a guest westerner who has never lived in a community so unlike his own, each individual's reactions vary from delight to horror, from fear to happiness, and from boredom to an explosion of nervous energy. One of the surest signs of culture strain is when my guest wants to spend more and more time talking to me or lying in their hammock and less and less time talking to our Amazonian hosts.

Sometimes these folks feel that they are prepared for their trip because they have watched documentaries about tribal societies. But they soon learn that a documentary reveals nothing more than the minuscule list of items that attracted the attention of the filmmaker. The life that goes on behind the camera is much richer than the odorless, temperature-neutral visual and aural snippets collected on screen. A culture is a living thing that bears the accretions of generations of lives, of our suffering, triumph, defeat, of thousands of days of boring routine and the past interactions of people long dead. You cannot even begin to understand another culture or language without doing a lot of living and speaking.

When I am alone among the Pirahãs, or I am just with my own family with them, I become nearly bored to death at times. I do my research, eat my fish, walk around, talk a bit, and hit the hammock early. There is no one to discuss my culture with. No fancy restaurants. No wine. Just water, fish, manioc meal, popcorn, coffee, canned meat and a bit of exercise.

On my loneliest days I watch the Pirahãs fish, play tag, make arrows, nurse babies, tickle each other, and in all of their activities I wonder how they can tolerate the sheer monotony of doing the same thing over and over and over again, day in and day out, until they die.

By the same token, when they interrogate me about my life in the US, they find it stupefyingly uninteresting and wholly devoid of life. How can there be pleasure where people eat only bland food that someone else gets for them?

Each of us on occasion sees the other, perhaps even ourselves, leading the life of Sisyphus, condemned to day upon day of hard labor without accomplishment. But what did Camus actually say about this? He said, surprisingly, that the life of Sisyphus can be a happy one. He has his daily goal and he reaches it. Judged by the day, Sisyphus's efforts succeed, as did mine working for the post office. Hour by hour, he makes measurable progress toward his objective.

The Pirahãs have grasped this. Like angst-free, realized existentialists, they embrace the accomplishments of each day and find meaning in their lives without worrying about their children's future or what posterity will think of them. They stare into the eyes of death without blinking and live their physically demanding lives almost constantly laughing and smiling. Their happiness and their lack of worry, the absence of preoccupation with the past, their refusal to fear the future, these things have shaped their language so as to exclude talk about the remote times, whether future or past, and to eschew numbers and counting, and to avoid complex sentences, because only people, things, and events for which there is direct evidence can participate, placing the burden of their communication on their stories rather than their sentences. They reject career goals and enjoy each day as it comes.

Maybe I am wrong about the reasons for the Pirahãs' happiness. There is so much I do not, nor ever will, understand about them. But I know what I see and what I hear. What I see are smiles, touching, and laughter that I hear nowhere else and that I have never encountered on this scale among any of the other twenty or so Amazonian groups I have visited over the years. And what I hear is a language without numbers, without talk of the future or the past, with a concern for the immediate and the useful that is reflected even in its verbs. I see a society of intimates whose affection for one another, valuation of the immediacy of experience, and desire to communicate well in a special society have shaped their very grammar.

Their happiness has seeped into their language. It has become a grammar of happiness. Each language brings its speakers happiness. We all possess grammars of happiness – our identities and our cultural cloaks.

The Bible is the first source I am aware of that recognizes language is a tool. As we know, in the Genesis story of the Tower of Babel, God is upset by the accomplishments of his creation, humans, because these creatures use language to work together to construct a tower. All this tower-building irritates God. He sees the humans' ability to work together on their tower as the direct result of their ability to talk to one another. Genesis thus takes the linguistic principle of 'You talk like who you talk with' and reverses it so that it becomes 'You are not going to talk like the ones you were talking with.' According to Genesis, God punished mankind by 'confusing[ing] the language of the whole earth' and, apparently, giving every human a different language – or every family or every clan (Genesis isn't clear on this and it doesn't really matter). God believed that people could work together unless they spoke the same language. He intended his diversification of human languages to be a punishment, making it impossible for us to cooperate as a species.

Well, the punishment backfired. The result is the fact that humans speak roughly 7,000 languages today instead of one, and so we have 7,000 highly sophisticated cognitive tool kits to draw from as a species. It is not merely the case that each language is only a tool for its speakers, but that it is a potential tool for all humans.

Languages are repositories of art, science, and all the knowledge and skills learned by humans. The more of these there are, the wealthier the race is.

According to Genesis, many languages came out of one. To borrow writer David Foster Wallace's reversal of the American latinate motto, the likely history of our cultures and languages is summed up as *e unibus pluram*, out of one, many.

The idea that all our current languages trace back to an original human tongue is scientifically plausible. Even likely. But there is no need to attribute current linguistic diversity to God. Natural forces always ensure linguistic mitosis – each language will divide to form two whenever the right conditions are found.

English used to be a dialect of German, spoken by the Saxons that left Germany to settle in the cold, wet land of Britain in the fifth century AD. For the first years subsequent to the Saxons' settlement in England, their language was identical to the language of their relatives in Saxony. But now English and German are distinct, mutually unintelligible languages.

Why did English stop being German? For the exact reasons that you talk more like the friends you grew up with than your own parents; for the reason that you talk more like members of your economic class than another. 'You talk like who you talk with.' When the English Saxons stopped talking to the German Saxons, natural adaptations, innovations, and effects of new linguistic and cultural surroundings led to regular change over time that culminated in the splitting of Saxon into

German Saxon and English. As the Mexican writer Carlos Fuentes said, 'So languages live, like bread and love, by being shared with others.'

Is the birth of English and the loss of our Saxon language a good thing or a bad thing? To put it another way, does anyone believe that England has deteriorated culturally as a result of developing a new language? Hardly. Language mitosis is not language loss. It is adaptation of the tool to fit the local environment.

But when a language dies, we all lose. When a language is born, we gain – the natural specialization of our species. Learning a second language can help to form partnerships, but for the species as a whole, depth and breadth of knowledge are best served by language diversity. The story of Babel is in fact a reminder of the beauty of this diversity, of human cognitive flexibility – we are not, after all, dogs that all bark alike! We are *Homo sapiens*. We are *Homo loquax*.

One prominent linguist has claimed that the loss of a single language is worse than a bomb dropped on the Louvre. Like many thousands of people, I have walked through that museum that was established by the leaders of the French Revolution in the old palace of Louis XIV on the bank of the river Seine and built up by Napoleon with loot plundered from the far reaches of his empire. The Louvre, like museums throughout the world, is a repository of art, knowledge, and accomplishment. And yet the French possess a repository that dwarfs the Louvre – their own beautiful language and the culture that grew alongside and nourishes it. Languages die as the cultures which sustain them die or change radically.

In the sixteenth century, marauders from Extremadura and other parts of Spain burned the libraries of the Aztecs and the Mayas of Mexico. These civilizations not only produced great art, buildings, and societies, but they gave us new foods and understandings of the 'New World.' An example of the kind of knowledge that was lost from so many American civilizations, such as the Maya, the Aztec, and the Inca, comes from the European-Americans' contact with the Patuxet man Tisquantum, whom we are more familiar with by the pronunciation Squanto.

Squanto helped the pilgrim-immigrants through their first New England winter. He and his people taught them how to plant, hunt, fish, and survive in this hostile new world. He tried to learn the culture and language of the new settlers and to serve as a bridge between them, his own people, and other communities indigenous to the Americas. He failed. His efforts were distrusted and he himself was poorly understood when he died. It is believed that his death was caused by poisoning by other indigenous peoples who believed that he had merely helped outsiders to take over Indian lands.

Tisquantum and his people died and the pilgrims took his village and renamed it Plymouth. It is a noble task to try to understand others and to have them understand you. But it is never an easy one.

In this book I have tried to explore peoples' languages, from the structure of their stories to the formation of their words. I have looked at human anatomy and human development. I have considered linguistic complexity and its connection to the general property of hierarchical organization that characterizes galaxies, atoms, and human languages. I have shown that animals can perform many of the cognitive tasks common to humans, though not always as well. In the book's journey from the Amazon to Europe, from the mind to the tongue, I have sought out evidence of the occasionally invisible, but always powerful hand of culture.

The point of this discussion was never to be 'right.' It would be naive to expect that any one study could reverse the course of the thoughts of the majority about the nature of human language. Rather it is to engage in debate and dialog, with the hope that others may direct their research and thoughts to the investigation of the connection between language and human values and society that motivated the first generation of American linguists, the descriptivists.

Language reveals the engine of our souls, our mind. It illuminates us and energizes us. We share it with all we meet. It is the cognitive fire of human life.

# EXTRACT FROM PLATO'S *MENO*

**Socrates:** *It will be no easy matter, but I will try to please you to the utmost of my power. Suppose that you call one of your numerous attendants, that I may demonstrate on him.*

**Meno:** *Certainly. Come hither, boy.*

**Socrates:** *He is Greek, and speaks Greek, does he not?*

**Meno:** *Yes, indeed; he was born in the house.*

**Socrates:** *Attend now to the questions which I ask him, and observe whether he learns of me or only remembers.*

**Meno:** *I will.*

**Socrates:** *Tell me, boy, do you know that a figure like this is a square?*

**Boy:** *I do.*

**Socrates:** *And you know that a square figure has these four lines equal?*

**Boy:** *Certainly.*

**Socrates:** *And these lines which I have drawn through the middle of the square are also equal?*

**Boy:** *Yes.*

**Socrates:** *A square may be of any size?*

**Boy:** *Certainly.*

**Socrates:** *And if one side of the figure be of two feet, and the other side be of two feet, how much will the whole be? Let me explain: if in one direction the space was of two feet, and in other direction of one foot, the whole would be of two feet taken once?*

**Boy:** *Yes.*

**Socrates:** *But since this side is also of two feet, there are twice two feet?*

**Boy:** *There are.*

**Socrates:** *Then the square is of twice two feet?*

**Boy:** *Yes.*

**Socrates:** *And how many are twice two feet? Count and tell me.*

**Boy:** *Four, Socrates.*

**Socrates:** *And might there not be another square twice as large as this, and having like this the lines equal?*

**Boy:** *Yes.*

**Socrates:** *And of how many feet will that be?*

**Boy:** *Of eight feet.*

**Socrates:** *And now try and tell me the length of the line which forms the side of that double square: this is two feet – what will that be?*

**Boy:** *Clearly, Socrates, it will be double.*

**Socrates:** *Do you observe, Meno, that I am not teaching the boy anything, but only asking him questions; and now he fancies that he knows how long a line is necessary in order to produce a figure of eight square feet, does he not?*

**Meno:** *Yes.*

**Socrates:** *And does he really know?*

**Meno:** *Certainly not.*

**Socrates:** *He only guesses that because the square is double, the line is double.*

**Meno:** *True.*

**Socrates:** *Observe him while he recalls the steps in regular order.* (**To the Boy:**) *Tell me, boy, do you assert that a double space comes from a double line? Remember that I am not speaking of an oblong, but of a figure equal in every way, and twice the size of this – that is to say of eight feet; and I want to know whether you still say that a double square comes from a double line?*

**Bo:y** *Yes.*

**Socrates:** *But does not this line become doubled if we add another such line here?*

**Boy:** *Certainly.*

**Socrates:** *And four such lines will make a space containing eight feet?*

**Boy:** *Yes.*

**Socrates:** *Let us describe such a figure: Would you not say that this is the figure of eight feet?*

**Boy:** *Yes.*

**Socrates:** *And are there not these four divisions in the figure, each of which is equal to the figure of four feet?*

**Boy:** *True.*

**Socrates:** *And is not that four times four?*

**Boy:** *Certainly.*

**Socrates:** *And four times is not double?*

**Boy:** *No, indeed.*

**Socrates:** *But how much?*

**Boy:** *Four times as much.*

**Socrates:** *Therefore the double line, boy, has given a space, not twice, but four times as much.*

**Boy:** *True.*

**Socrates:** *Four times four are sixteen, are they not?*

**Boy:** *Yes.*

**Socrates:** *What line would give you a space of eight feet, as this gives one of sixteen feet, do you see?*

**Boy:** *Yes.*

**Socrates:** *And the space of four feet is made from this half line?*

**Boy:** *Yes.*

**Socrates:** *Good; and is not a space of eight feet twice the size of this, and half the size of the other?*

**Boy:** *Certainly.*

**Socrates:** *Such a space, then, will be made out of a line greater than this one, and less than that one?*

**Boy:** *Yes; I think so.*

**Socrates:** *Very good; I like to hear you say what you think. And now tell me, is not this a line of two feet and that of four?*

**Boy:** *Yes.*

**Socrates:** *Then the line which forms the side of eight feet ought to be more than this line of two feet, and less than the other of four feet?*

**Boy:** *It ought.*

**Socrates:** *Try and see if you can tell me how much it will be.*

**Boy:** *Three feet.*

**Socrates:** *Then if we add a half to this line of two, that will be the line of three. Here are two and there is one; and on the other side, here are two also and there is one: and that makes the figure of which you speak?*

**Boy:** *Yes.*

**Socrates:** *But if there are three feet this way and three feet that way, the whole space will be three times three feet?*

**Boy:** *That is evident.*

**Socrates:** *And how much are three times three feet?*

**Boy:** *Nine.*

**Socrates:** *And how much is the double of four?*

**Boy:** *Eight.*

**Socrates:** *Then the figure of eight is not made out of a three?*

**Boy:** *No.*

**Socrates:** *But from what line? Tell me exactly; and if you would rather not reckon, try and show me the line.*

**Boy:** *Indeed, Socrates, I do not know.*

**Socrates:** *Do you see, Meno, what advances he has made in his power of recollection? He did not know at first, and he does not know now, what is the side of a figure of eight feet: but then he thought that he knew, and answered confidently as if he knew, and had no difficulty; now he has a difficulty, and neither knows nor fancies that he knows.*

# ACKNOWLEDGMENTS

Many people have read portions of this manuscript and offered valuable advice. I would like to thank Caleb Everett, Tom Wolfe, John Davey, Edward Kastenmeier, Ted Gibson, Steve Piantadosi, Miguel Oliveira Jr, Andy Weeks, Sascha Griffiths, Jan Wohlgemuth, Paul Hopper, Linda Everett, Sindhu Palaniappan, David Brumble, Geoff Pullum, Sally Thomason, Mitchell Mattox, Max Brockman, Luis Miguel Rojas, Heidi Keller, Khaled Hakami, Andrew Franklin, Jim Swindler, Michael Frank, Malte Henk, Robert van Valin Jr, Jeanette Sakel, Shalom Lappin, Paul Postal, Philip Lieberman, Mark Liberman, and the Royal Institution, among others, for opportunities to discuss my ideas and for many detailed comments.

# LIST OF FIGURES

# SUGGESTED READING

Anderson, Stephen R. and David W. Lightfoot, *The Language Organ: Linguistics as Cognitive Physiology* (Cambridge: Cambridge University Press, 2002)

Aristotle, *The Politics*, ed. Trevor J. Saunders (Harmondsworth and New York: Penguin Classics, revised edn, 1981)

Austin, J. L., *How to Do Things with Words: Second Edition* (Cambridge, MA: Harvard University Press, 1975)

Berlin, Brent and Paul Kay, *Basic Color Terms: Their Universality and Evolution* (Stanford, CA: Center for the Study of Language and Information, 1999)

Boas, Franz, *Race, Language and Culture* (Chicago: University of Chicago Press, 1995)

Boyd, Robert and Peter J. Richerson, *Culture and the Evolutionary Process* (Chicago: University of Chicago Press, 1988)

Boyd, Robert and Peter J. Richerson, *The Origin and Evolution of Cultures* (New York: Oxford University Press, 2005)

Bruner, Jerome, *Acts of Meaning: Four Lectures on Mind and Culture* (Cambridge, MA: Harvard University Press, 1992)

Bruner, Jerome, *The Culture of Education* (Cambridge, MA: Harvard University Press, 1996)

Burton, Frances D., *Fire: The Spark That Ignited Human Evolution* (Albuquerque: University of New Mexico Press, 2009)

Carey, Susan, *The Origin of Concepts* (Oxford and New York: Oxford University Press, 2009)

Chomsky, Noam, *Syntactic Structures* (The Hague: Mouton de Gruyter, 1957)

Chomsky, Noam, *Aspects of the Theory of Syntax* (Cambridge, MA: The MIT Press, 1969)

Chomsky, Noam, *The Minimalist Program* (Cambridge, MA: The MIT Press, 1995)

Churchland, Patricia Smith, *Neurophilosophy: Toward a Unified Science of the Mind-Brain* (Cambridge, MA: The MIT Press, 1989)

Churchland, Paul M., *The Engine of Reason, the Seat of the Soul: A Philosophical Journey into the Brain* (Cambridge, MA: The MIT Press, 1996)

Corballis, Michael C., *The Recursive Mind: The Origins of Human Language, Thought, and Civilization* (Princeton: Princeton University Press, 2011)

Cowie, Fiona, *What's Within? Nativism Reconsidered* (New York: Oxford University Press, 1999)

Culicover, Peter W. and Ray Jackendoff, *Simpler Syntax* (Oxford and New York: Oxford University Press, 2005)

Culicover, Peter W., *Syntactic Nuts: Hard Cases in Syntax* (Oxford University Press, 1999)

Elman, Jeffrey L., Elizabeth A. Bates, Mark H. Johnson, Annette Karmiloff-Smith, Domenico Parisi, and Kim Plunkett, *Rethinking Innateness: A Connectionist Perspective on Development* (Cambridge, MA: The MIT Press, 1997)

Enfield, Nick (ed.), *Ethnosyntax: Explorations in Grammar and Culture* (Oxford and New York: Oxford University Press, 2002)

Enfield, Nicholas J. and Stephen C. Levinson, *The Roots of Human Sociality: Culture, Cognition, and Interaction* (Oxford: Berg Publishers, 2006)

Gazdar, Gerald, Ewan Klein, Geoffrey K. Pullum, and Ivan Sag, *Generalized Phrase Structure Grammar* (Cambridge, MA: Harvard University Press, 1985)

Givon, Talmy, *On Understanding Grammar* (New York: Academic Press, 1987)

Goldberg, Adele E., *Constructions at Work: The Nature of Generalization in Language* (New York and Oxford: Oxford University Press, 2006)

Goodman, Nelson, *Of Mind and Other Matters* (Cambridge, MA: Harvard University Press, 1984)

Greenberg, Joseph H. (ed.), *Universals of Language* (Cambridge, MA: The MIT Press, 1963)

Hrdy, Sarah B., *Mothers and Others: The Evolutionary Origins of Mutual Understanding* (Cambridge, MA: Belknap Press, 2009)

Hume, David, *An Enquiry Concerning Human Understanding* (1748; Oxford and New York: Oxford University Press, 1999)

Hurford, James R., *The Origins of Meaning: Language in the Light of Evolution* (Oxford: Oxford University Press, 2007)

Lee, Namhee, Lisa Mikesell, Anna Dina L. Joaquin, Andrea W. Mates, and John H. Schumann, *The Interactional Instinct: The Evolution and Acquisition of Language* (New York: Oxford University Press, 2009)

Levi-Strauss, Claude, *Tristes Tropiques*, trans. John and Doreen Weightman (Harmondsworth and New York: Penguin, 1992)

Lewis, David K., *Convention: A Philosophical Study* (Oxford: Blackwell, 2002)

Lieberman, Philip, *Eve Spoke: Human Language and Human Evolution* (New York: W. W. Norton and Company, 1998)

Lieberman, Philip, *Toward an Evolutionary Biology of Language* (Cambridge, MA: Harvard University Press, 2006)

Locke, John, *An Essay Concerning Human Understanding* (1689; Oxford: Clarendon Press, 1979)

Perkins, Revere D., *Deixis, Grammar, and Culture* (Amsterdam and Philadelphia: John Benjamins, 1992)

Piaget, Jean, *The Language and Thought of the Child* (1926; Goldberg Press, 2008)

Pinker, Steven, *The Language Instinct: How the Mind Creates Language* (1994; New York: Harper Perennial Modern Classics, 2007)

Plato, *Cratylus* (Gloucester: Dodo Press, 2007)

Plato, *Meno* (Akasha Classics, 2009)

Pratt, Scott L., *Native Pragmatism: Rethinking the Roots of American Philosophy* (Bloomington, IN: Indiana University Press, 2002)

Pulvermüller, Friedemann, *The Neuroscience of Language: On Brain Circuits of Words and Serial Order* (Cambridge: Cambridge University Press, 2003)

Quine, Willard van Orman, *Word and Object* (Cambridge, MA: The MIT Press, 1964)

Richerson, Peter J. and Robert Boyd, *Not by Genes Alone: How Culture Transformed Human Evolution* (Chicago: University of Chicago Press, 2004)

Rorty, Richard, *Contingency, Irony, and Solidarity* (Cambridge: Cambridge University Press, 1989)

Sapir, Edward, *Language: An Introduction to the Study of Speech* (1921; Hard Press Publishing, 2010)

Searle, John R., *Speech Acts: An Essay in the Philosophy of Language* (Cambridge: Cambridge University Press, 1970)

Searle, John R., *Intentionality: An Essay in the Philosophy of Mind* (Cambridge: Cambridge University Press, 1983)

Searle, John R., *Mind, Language, and Society: Philosophy in the Real World* (New York: Basic Books, 2000)

Searle, John R., *Making the Social World: The Structure of Human Civilization* (New York: Oxford University Press, 2010)

Shannon, Claude E., *The Mathematical Theory of Communication* (Champaign, IL: University of Illinois Press, 1998)

Simon, Herbert A., *The Sciences of the Artificial* (Cambridge, MA: The MIT Press, 1969)

Skinner, B. F., *Science and Human Behavior* (New York: Free Press, 1965)

Tomasello, Michael, *The Cultural Origins of Human Cognition* (Cambridge, MA: Harvard University Press, 1999)

Tomasello, Michael, *Constructing a Language: A Usage-Based Theory of Language* (Cambridge, MA: Harvard University Press, 2003)

Tomasello, Michael, *Origins of Human Communication* (Cambridge, MA: The MIT Press, 2008)

Van Valin Jr, Robert D. and Randy J. LaPolla, *Syntax: Structure, Meaning, and Function* (Cambridge: Cambridge University Press, 1998)

Von Humboldt, Wilhelm, *On Language: On the Diversity of Human Language Construction and Its Influence on the Mental Development of the Human Species* (Cambridge: Cambridge University Press, 2nd edn, 2000)

Vygotsky, L. S., *Mind in Society: The Development of Higher Psychological Processes* (Cambridge, MA: Harvard University Press, 1978)

Whorf, Benjamin Lee, *Language, Thought, and Reality: Selected Writings of Benjamin Lee Whorf* (Eastford, CT: Martino Fine Books, 2011)

Wittgenstein, Ludwig, *Philosophical Investigations* (1953; Oxford: Wiley-Blackwell, 4th edn, 2009)

Wrangham, Richard, *Catching Fire: How Cooking Made Us Human* (New York: Basic Books, 2009)

# INDEX